T0305656

The Ultimate Student's Guide to Scientific Research

A scientific career is a long and winding journey. Many factors assemble to determine the trajectory and products of scientific inquiry. This book addresses the scientific career path comprehensively, ranging from how to build a strong foundational knowledge and skill base, to training programs, composing winning research proposals and grants, conducting excellent research, writing papers and reports, collaborative research programs, bridging academia and industry, career advancement, and more.

Features:

- Guides where, how, and what to study in undergraduate, post-baccalaureate, graduate, and post-graduate training programs and addresses important crossroads throughout the continuum of training and beyond.
- Highlights best practices, techniques, and nuances for creating a successful scientific career.
- Provides critical insights for traversing major milestones and decision-points in a scientific career and serves as a resource for reference throughout stages of advancement.
- Discusses alternative career opportunities for individuals trained in sciences.
- Offers students, researchers, and other scientists across all stages of their careers with accessible, engaging, and useful insider tips.

The Ultimate Student's Guide to Scientific Research

Samuel J.S. Rubin, PhD, MD
Stanford University School of Medicine

Nir Qvit, PhD, MBA
Faculty of Medicine, Bar-Ilan University

CRC Press
Taylor & Francis Group
Boca Raton London New York

CRC Press is an imprint of the
Taylor & Francis Group, an **informa** business

Front cover image: Artwork created by Mahesh Kumar Cinthakunta Sridhar.

First edition published 2024
by CRC Press
2385 Executive Center Drive, Suite 320, Boca Raton, FL 33431

and by CRC Press
4 Park Square, Milton Park, Abingdon, Oxon, OX14 4RN

CRC Press is an imprint of Taylor & Francis Group, LLC

Library of Congress Cataloguing-in-Publication Data
Names: Rubin, Samuel J. S., author. | Qvit, Nir, author.
Title: The ultimate scientist's guide to the galaxy / by Samuel J.S. Rubin, PhD, MD, Stanford University School of Medicine, Nir Qvit, PhD, MBA, Faculty of Medicine, Bar-Ilan University.
Description: Boca Raton : CRC Press, 2024. |
Identifiers: LCCN 2024003274 (print) | LCCN 2024003275 (ebook) | ISBN 9781032293875 (hardback) | ISBN 9781032293820 (paperback) | ISBN 9781003301400 (ebook)
Subjects: LCSH: Scientists. | Science--Vocational guidance.
Classification: LCC Q147 .R83 2024 (print) | LCC Q147 (ebook) | DDC 502.3--dc23/eng/ 20240412
LC record available at https://lccn.loc.gov/2024003274
LC ebook record available at https://lccn.loc.gov/2024003275

ISBN: 978-1-032-29387-5 (hbk)
ISBN: 978-1-032-29382-0 (pbk)
ISBN: 978-1-003-30140-0 (ebk)

DOI: 10.1201/9781003301400

Typeset in Times
by MPS Limited, Dehradun

Dedication

To all our mentors and mentees – past, present, and future – this is inspired by you, this is for you, and this is in your honor. This is especially in the loving memory of Jonathan C. Wright, PhD.

Contents

About the Authors

Samuel J. S. Rubin, PhD, MD

Dr. "Yoni" Samuel J. S. Rubin completed his PhD in immunology at Stanford University School of Medicine, followed by postdoctoral training, MD, and postgraduate clinical training at Stanford. His research focuses on better understanding the molecular and cellular mechanisms underlying chronic immune-mediated diseases and using this knowledge to develop safer, cheaper, and more effective tools for detecting and treating illness worldwide. In the field of mucosal immunology and immune cell trafficking, Dr. Rubin's findings have inspired the development of novel blood-based methods for detection of gastroenterological conditions. His work has also led to the development of precision medicine biologics for the treatment of chronic auto-inflammatory conditions. Dr. Rubin is especially appreciative for the circuitous and often unexpected path that continues to define his career. Especially influential were the many instructors and mentors who inspired his dedication to teaching the student and shaped his past and ongoing endeavors. He has published *Peptides and Peptidomimetics: From Bench to Bedside*, served as a guest editor for *Current Topics in Medicinal Chemistry*, and continues to serve as a reviewer for numerous scientific journals. Amongst other awards and recognitions, Dr. Rubin received the United States National Science Foundation Graduate Research Fellowship and the Hugh McDevitt Prize in Immunology.

Nir Qvit, PhD, MBA

Dr. Nir Qvit completed his doctorate in organic chemistry at the Hebrew University in 2008. His graduate work focused on developing strategies for synthesis of small molecules, peptides, and peptidomimetics (modified peptides) for various therapeutic applications. Dr. Qvit completed postdoctoral training and worked as a Senior Scientist in Chemical and Systems Biology at Stanford University. In 2017, Dr. Qvit became a senior lecturer and principal investigator of the multi-national and inter-disciplinary Laboratory for Chemical Biology of Protein-Protein Interactions for Drug Discovery in the Azrieli Faculty of Medicine at Bar Ilan University. His current research focuses on the development of novel tools to regulate protein-protein interactions. Dr. Qvit uses a rational approach to design and develop short peptides and peptidomimetics derived from protein regulatory domains to modulate their function in a selective manner. He has published *Peptides and Peptidomimetics: From Bench to Bedside*, served as a guest editor for *Current Topics in Medicinal Chemistry*, and continues to serve as a reviewer for many scientific journals, including the *Journal of Medicinal Chemistry* and the *Journal of Peptide Science*.

1 Introduction

Samuel J. S. Rubin and Nir Qvit

WHAT OR WHO IS A "SCIENTIST"?

Hearing the term "scientist" can mean vastly different things to different people. Some envision the classical investigator working in a laboratory, experimenting with Petri dishes or flasks, and aiming to discover a new subatomic particle, a new chemical reaction, or a new disease treatment. Others think of environmental scientists studying the climate or earth systems, mathematicians and computer scientists conceiving new theorems and algorithms, or clinical scientists and epidemiologists characterizing public health trends. Herein, we focus on the scientist as an investigator who utilizes what is now thought of as the scientific method to experiment, probe, and advance knowledge grounded in fundamental physics, chemistry, and/or biology – which may assume any of the aforementioned manifestations and more. Simply put, a scientist experiments in a medium, making discoveries of new knowledge – sometimes fundamental and other times applied. In other words, a scientist is an artist whose medium is asking questions. A scientist plays fun, exciting, challenging, confusing, frustrating, and marvelous roles.

The scientific method can be broadly understood as an iterative process founded in rigorous skepticism that includes an ongoing repetitive sequence of observation, hypothesis generation, experimentation, data analysis, and interpretation. To accomplish this feat, successful scientists often work in teams, collaborate across disciplines, and spend years honing their technical skills and capacity for critical thinking. In these ways, a scientist always remains a student. Scientific knowledge is, or should be, an ever-expanding, self-correcting, replicating body of work based on publicly available and clearly presented unequivocal evidence, or data. But just as every field is inevitably permeated by money, politics, and individual ego, so too is science. Whether funding sources or political alignments influence data reported or conclusions drawn, journals exert influence over the flow of information, or scientists themselves fail to present their findings in an accessible or accurate format – adoption of scientific knowledge is limited by susceptibility to individual trust and by pre-existing belief. Thus, scientific knowledge is tremendously dependent upon a sacred vow to fundamental objectivity and transparency in use of the scientific method which must be preserved.

GUIDING VALUES

Becoming and excelling as a scientist is a tremendously challenging endeavor, requiring immense stamina, passion, and curiosity. Much like the experiments

DOI: 10.1201/9781003301400-1

conducted, scientific training and career advancement can often involve substantial trial and error, as the profession is largely decentralized and heterogeneous. Having experienced this first hand, we decided to compile this work as a guide and reference for potential scientists, aspiring scientists, scientists in training, scientists seeking advice, and anyone else seeking a window into the scientific world. Equally and perhaps more important is our philosophy of paying it forward. Though we both navigated our paths by significant trial and error, we also benefited greatly from the advice, guidance, and counsel of others who took the time and effort to share their experiences with us – never in exchange for compensation but solely on the condition that we continue to pass on what we learn to others.

As you will see throughout this book, paying it forward is one of our most important guiding values for being a successful scientist and a responsible individual. This also translates into treating others the way you wish to be treated, which will get you much farther in the world than alternative approaches. Other requirements for being a successful scientific investigator are an unquenchable thirst for continual learning, insatiable curiosity, and a healthy dose of skepticism. These attributes are key for maintaining inspiration, gaining new knowledge, making novel discoveries, and avoiding all too common pitfalls of cognitive bias.

Other important practices are seeking different perspectives and learning to teach yourself anything you may need. Actively pursuing distinct perspectives will broaden your options and capabilities, teach you things that you did not realize you needed to learn, allow you to ask better questions, and help avoid different forms of cognitive bias. Learning to teach yourself anything will empower you to excel, to be agile, and to avoid dependence on any single entity. Teaching yourself can assume a variety of forms, ranging from reading a textbook, searching the Internet, seeking out and consulting with experts, taking a course, or performing an experiment. Building this dexterity and initiative is invaluable for the ever-evolving pursuit of science.

Learning from failure is immensely helpful in streamlining subsequent efforts and is central to the process of trial and error. However, learning from the error of others or by observing something done in a manner incongruent with scientific standards or your own values is equally or perhaps more critical than growth from your own trial and error because with this external awareness you may be able to avoid making crucial mistakes in the first place. Paying attention to others' failures, their challenges, how someone accomplishes a task even successfully but in a manner that may be right for them but wrong for you, or simply identifying that a particular approach feels wrong to you, are all important for heading off your own missteps before they occur and choosing an appropriate path forward. Some of the most valuable lessons we have learned about how we want to conduct experiments, run a lab, write a paper, etc. have originated from observing the way others approach these tasks – both successfully and unsuccessfully – and using those observations and our values to hone our own approaches.

WHAT WE HOPE YOU WILL GAIN FROM THIS BOOK

Our hope is that this book will serve as a guide, a reference, and a window into the scientist and their journey through the scientific world. Whether considering this life career, seeking advice on the path, or looking for an understanding of what a scientist encounters, we hope this book will provide you with useful information. We include many lessons that we wish we understood earlier. We present a variety of anecdotes and examples, some personal and others that we witnessed or learned from. We changed the names of the individuals involved to protect their privacy. We will be candid – not to dissuade anyone from approaching the marvelous role of the scientist that of course comes with its own set of obstacles just as every field and profession – but rather to give those interested a head start and a better chance at fulfillment and excellence, having known what lies ahead and been prepared to navigate the challenges.

2 The History of Science

Nir Qvit and Samuel J. S. Rubin

HISTORY OF SCIENCE IN EARLY CULTURES

Ancient societies did not have modern laboratory-based science, and their work is often dismissed as pseudoscience today. However, even in the Stone Age, people used science to survive. Archaeological evidence indicates astronomical knowledge development in preliterate societies before writing systems. Prehistoric people used celestial bodies to keep time beyond a single day and night. For example, the moon was one of the earliest timepieces, as its face changed nightly throughout seasons, allowing it to be relied upon for time keeping. There is substantial evidence that lunar timekeeping was observed as early as 35,000 years ago and that humankind has kept calendars dating back to the late Paleolithic era – around 32,000 years ago. Yet, since in prehistoric times, knowledge was passed from generation to generation orally, we have very limited details.

Hunting was the original and main occupation of prehistoric people, and ancient humans used complex hunting techniques to ambush and kill large animals. The development and promotion of sustainable hunting depended on science. During the Stone Age, hunters observed the habits of animals they used to hunt, as seen in paintings in many caves around the world. This includes the Cave of Niaux, located in the Niaux commune in southwestern France. In many such places, wall paintings demonstrate movements and even animal anatomy derived by methods akin to the modern scientific process.

Agriculture is considered the next occupation after hunting and demanded technological development (*e.g.*, tools) as well as the need for a more accurate calendar that guides farmers on when to sow and reap crops. This led to knowledge accumulation and contributed to scientific progress. Based on the development of science and technology, people were able to produce enough crops to feed themselves and trade the extra food they produced. Six distinct cradles of civilization that depended upon agriculture for sustenance emerged independently – four in the Old World, including Mesopotamia, Indus Valley, China, and Egypt, as well as two in the New World, including Peru and Mexico. In all these societies, cities were built, technological innovations were made, language was invented, and complex social orders came into being. Historians generally acknowledge that all the major ancient civilizations developed in much the same way, despite regional and climatic differences. Farming that made settlements possible required reliable water supplies. Therefore, early villages grew along rivers, becoming permanent

DOI: 10.1201/9781003301400-2

settlements with wood, brick, and stone structures. These villages later developed into the first cities.

Sumerians formed the first human civilization in Mesopotamia (Greek: "Land between the Rivers") around 5,500 years ago. They are credited with many advances in technology, such as irrigation systems, the wheel, the plow, and writing (cuneiform). Sumerians also had an interest in science including astronomy and created a calendar adjusted to the phases of the moon. In addition, they created a system of degrees, minutes, and seconds.

Other civilizations also studied science and developed similar technology. For example, the Egyptians also invented the plow, used copper materials, invented writing (hieroglyphic), and the first 365-day calendar. The Chinese were skilled chemists, and they performed sophisticated reactions, such as distilling alcohol and extracting copper. In addition, they developed gunpowder by blending charcoal, sulfur, and potassium nitrate.

In many of these civilizations, people studied the moon, sun, and planet movements to calculate seasons, establishing a ritual calendar. Therefore, it is not surprising that astronomy is ranked as one of the oldest sciences and the first natural science developed by early civilizations, as knowledge of the stars proved essential to running a complex agricultural society. Ancient civilizations also used mathematics for practical applications in various fields, such as astronomy (*e.g.*, measuring the motion of celestial bodies), engineering (*e.g.*, construction of monumental structures), agriculture (*e.g.*, measuring land and calculating field area), and timekeeping. Texts from about 1,700 BCE demonstrate remarkable mathematical elegance, and Babylonian mathematicians understood the Pythagorean relationship. However, there are fewer examples of these early groups studying beyond practical calculations applied to their real world needs.

In all these early civilizations, humans developed methods to describe and harness nature. However, to understand nature was the function of religion and magic, not reason. The Greeks were the first to explore beyond description and arrive at reasonable explanations for natural phenomena, which did not involve the divine will.

CLASSICAL ANTIQUITY

Based on the contributions of ancient civilizations across fields, ancient Greeks attempted to explain world events by invoking natural causes. They are often called the first scientists or natural philosophers. Numerous inventions and discoveries are attributed to ancient Greek scientists in various fields, such as astronomy, geography, and mathematics. It was a period characterized by a flourishing and confident Greek culture in many aspects. This included politics, philosophy, science, architecture, theater, athletic games, and military strength that formed a legacy with unparalleled influence on Western civilization.

The ancient Greeks used science and logic to begin understanding events happening in the world around them. They made many advancements in science and technology. Thales, one of the first Greek mathematicians, focused on geometry and observed many key relationships (such as Thale's theorem) concerning shapes, angles, and lines. Pythagoras defined the Pythagorean Theorem, which describes the sides of a right triangle and is still regularly utilized to this day across numerous fields. Euclid, regarded as the most significant Greek mathematician, wrote several books on the subject of geometry. Euclid's compilation *Elements* includes 13 books on geometry, which served as the world's main text for almost two millennia. Archimedes is also well known for many discoveries such as the relationship between the surface and volume of a sphere and its circumscribed cylinder, formulation of the hydrostatic principle (Archimedes' principle), and a strategy to move water (Archimedes' screw).

Astronomy was critical for navigation, understanding and regulating agriculture, and creating an accurate calendar. Development of the field by ancient Greek astronomers is considered a major phase in the history of astronomy. The Greeks introduced mathematics into astronomy, seeking a geometrical model for celestial phenomena. The Greek astronomer Aristarchus believed the sun was the center of the solar system. He also placed the planets in the correct order from the sun. Eratosthenes was the first to calculate the Earth's circumference accurately. They even developed a device called the Antikythera mechanism, which is considered the world's first analog computer. This device was used for calculating movements of the planets. The Greeks also applied their learning to many other practical inventions, such as the watermill, alarm clock, central heating, and crane, etc.

Ancient Greek society also made its mark in many other fields, such as physics. Their study was the first to be led by intellectual pursuit in a controlled manner, which is standard practice today. The Greeks also studied medicine as a scientific way to cure illnesses and diseases. Hippocrates, the most famous Greek doctor and sometimes considered to be the forebearer of Western medicine, taught that diseases had natural causes and could sometimes be cured by natural means. Indeed, Greek doctors studied sick people and devised treatments based on their symptoms. The Greeks likewise developed an encyclopedic range of herbal medicines. Many medical students still take the Hippocratic Oath upon graduation, although there are several antiquated portions that in many cases are modified and updated.

The ancient Greeks also explored biological processes and living organisms. The term "biology" is derived from the Greek terms "bios" ("life") and "logia" ("study of"). Aristotle studied animals in detail and made what is considered a record of the first systematic and comprehensive study of animals. He also noted his observations in a book titled *History of Animals*. Aristotle's work laid the foundations of zoology and botany by classifying animals and plants according to their different characteristics. Darwin considered Aristotle the most significant early contributor to biological thought. In an 1882 letter, Darwin wrote that "Linnaeus and Cuvier are my two gods, though in very different ways, but they were mere schoolboys to old Aristotle" (Gotthelf, 1999).

THE RISE OF MODERN SCIENCE

The scientific revolution began in Europe in the second half of the Renaissance period, with the 1543 publication by Nicolaus Copernicus entitled *De revolutionibus orbium coelestium* (on revolutions of heavenly spheres) and continuing through the late 18th century. At that time, a series of events changed scholarly thought, leading to the birth of modern science. During the scientific revolution, science became an autonomous discipline, distinct from philosophy and technology. Some scholars claim that at the end of this period, science replaced Christianity as the focal point of European civilization. The scientific revolution generated a new view of science, bringing about the following transformations: an emphasis on abstract reasoning; focus on quantitative description of nature; understanding of how nature works; the view of nature as a machine rather than an organism; and the development of an experimental scientific method.

A few of the key people and discoveries in this area include the astronomer Nicolas Copernicus, who demonstrated that Earth revolves around the sun and that the Earth is not the center of the universe, an idea that was embedded in the European conscience, which challenged societal values of the time. Galileo Galilei was an astronomer and physicist who was placed under house arrest for much of his life due to views that did not align with the Church; building on Copernicus' work, Galilei provided scientific insights that laid the foundation for many future scientists, sparking the field of modern astronomy. He was the first to use a refractory telescope to make significant astronomical discoveries. Galilei observed the moon, Venus phases, moons around Jupiter, sunspots, and seemingly countless individual stars that make up the Milky Way Galaxy. Galilei is often credited with the origin of modern physics for his role in the scientific revolution. His pioneering work on body motions was a precursor to Isaac Newton's classical mechanics. Later, Newton published *Philosophiæ Naturalis Principia Mathematica* (Mathematical Principles of Natural Philosophy) in 1687, which provided the foundations for classical mechanics, establishing the laws of motion and gravitation that revolutionized science.

In the early 1600s, the astronomer and physicist Johannes Kepler published Kepler's laws of planetary motion, which concern the elliptical motion of planets around the sun. Kepler's laws became part of the scientific revolution's central dogma and have been utilized in the development of numerous transformative technologies such as satellites and rockets.

Andreas Vesalius was an anatomist who published groundbreaking work on blood circulation in the 1500s. Based on that foundation, William Harvey demonstrated how blood circulates in the body and explained how the heart propels blood through the body, a transformational discovery for the time.

Francis Bacon was a politician and philosopher in the late 1500s and early 1600s who proposed a focus on observation and reasoning in the scientific method. He developed the scientific methodology, for which he is known as the creator of empiricism. Therefore, rigorous experimentation was to be used to prove or disprove hypotheses in the quest for understanding the universe.

SUMMARY

Methodical study of the physical, natural, and social worlds' behavior through experimentation and observation is understood to define modern science. Science tests theories against evidence. The study of science is one of the oldest and most significant academic disciplines, as it allows us to understand the world around us.

Moreover, science influences the countless decisions we make each day. Science is dynamic and built upon ideas and discoveries from previous generations. Therefore, the ability to recount the past and pass it on to future generations is critical. The history of science covers its development from ancient times to the present. Despite the limited examples presented in this chapter, it is particularly worthwhile to remember that science has no single origin. Rather, systematic methods emerged gradually and in parallel over thousands of years from a diversity of cultural traditions. What remains unchanged is the curiosity, imagination, and intelligence across generations and cultures.

REFERENCES

Casadevall, Arturo, and Ferric C. Fang. "(a) Historical Science." 4460-4464: American Society for Microbiology, 2015.
Daston, Lorraine. "History of Science." In *International Encyclopedia of the Social & Behavioral Sciences*, edited by Neil J. Smelser and Paul B. Baltes, 6842–6848. Oxford: Pergamon, 2001.
Gotthelf, Allan. "Darwin on Aristotle." *Journal of the History of Biology* 32, no. 1(1999): 3–30.
Hooykaas, R. "The Rise of Modern Science: When and Why?" *The British Journal for the History of Science* 20, no. 4 (1987): 453–473.
Kuhn, Thomas. "The History of Science." In *Philosophy, Science, and History*, 106–121. Routledge, 2014. Pages16, Taylor and Francis group. eBook ISBN9780203802458.
Lloyd, Geoffrey. "Science in Ancient Civilizations?" In *Ancient Worlds, Modern Reflections: Philosophical Perspectives on Greek and Chinese Science and Culture*, edited by Geoffrey Lloyd. Oxford University Press, 2004. 10.1093/0199270163.003.0002

3 Scientific Impacts

Nir Qvit and Samuel J. S. Rubin

EDUCATION

Science is an essential part of our education as it helps us comprehend the natural world around us. In learning about the world through scientific studies, we can understand fundamental principles that govern the universe, from the behavior of living organisms to the laws of physics. Moreover, science enables us to understand and address global challenges (*e.g.*, reducing greenhouse gas emissions, as well as pollution and poverty), and many of these challenges require deep scientific understanding. Scientific research also advances cutting-edge technologies, which improve quality and quantity of life (*e.g.*, life-saving medical treatments) and assist us in solving societal problems (*e.g.*, more efficient forms of energy).

First and foremost, science is based on evidence, and scientific methodology includes observation, experimentation, and analysis. Learning this method helps students develop critical thinking skills (*e.g.*, asking questions, seeking answers, and exploring the world around them), which they can apply to all aspects of their lives. In addition, by studying science, students learn to think creatively and develop creative ideas, since science education teaches us to think logically, analyze data, and make logical decisions based on evidence. In summary, science education is vital for our personal and societal well-being. Learning science helps us understand and solve problems, advances technology, and improves our quality of life.

MEDICINE

Science has allowed society to study and advance knowledge of the human body and the pathogenesis of diseases. This knowledge advances the development of more effective treatments and cures (*e.g.*, vaccines and antibiotics), saving millions of lives worldwide and significantly improving public health. Diseases such as smallpox and polio have been all but eradicated thanks to vaccinations. In addition, science has a vital role in the regulation of infectious diseases (*e.g.*, identifying and tracking COVID-19), as well as developing strategies to prevent and control outbreaks.

Without a doubt, science is one of the most significant pillars of medicine. Scientific research has led to fewer infections and lethal diseases, allowing physicians to cure or control them based on interdisciplinary inventions. Medical science has led to countless innovations, from new technologies and materials to effective treatments and therapies (*e.g.*, antibiotics, chemotherapy, immunotherapy,

DOI: 10.1201/9781003301400-3

organ transplantation, artificial limbs, hearing aids, X-rays, magnetic resonance imaging (MRI) machines, laparoscopic and robotic operating systems, etc.) that improve our quality and quantity of life. A side product of medical research innovation is its significant impact on the economy (*e.g.*, the biotech and pharmaceutical industries), generating millions of jobs and creating billions of dollars in economic growth activity. In summary, science is essential to our health as individuals and as a society. It helps prevent and treat illness, improve public health, and drive economic growth and innovation.

COMMUNICATION

Science plays a crucial role in communication, and communication is also an essential aspect of science. Telegrams, telephones, networking, fax, the Internet, laptops, and mobile phones are only some of the most significant contributions of science to communication, which have all transformed human life. As science is also of importance to the public, communicating it to the public can increase understanding of scientific concepts that can lead to better-informed decisions in numerous areas.

Effective communication is essential to ensure that scientific knowledge is disseminated accurately, efficiently, and ethically to a wider audience (*e.g.*, other scientists, funders, policymakers, consumers, and the public at large). Since in many cases science is a collaborative effort, effective communication is necessary to share ideas, methods, and data with colleagues to advance science. Efficient communication can help advance knowledge and understanding in a particular field and lead to new discoveries. Overall, communication is an essential component of the scientific process and improves our understanding of the world around us. Effective communication facilitates collaboration, promotes scientific research, and informs policy decisions, hopefully resulting in better-informed decisions.

AGRICULTURE

Agriculture is the second oldest and most fundamental occupation of humans since the stone age. Moreover, agriculture has undergone significant changes throughout history. Science shaped the development of modern agriculture, from traditional subsistence farming to modern agribusiness. Science has made farming easier and faster in many ways (*e.g.*, sowing seeds, harvesting, spraying fertilizers, irrigation, etc.) through improved technologies and machines. Agricultural science is an interdisciplinary field of food and fiber manufacture that includes many steps (*e.g.*, production, processing, and distribution) and deals with various disciplines (*e.g.*, plant breeding, genetics, soil science, entomology, plant pathology, and food science). Scientists have used plant selection and breeding techniques to improve crop production, yield, and quality, as well as enhance the nutritional content of crops. This helps meet the growing demand for food as the global population increases.

By adopting modern techniques (*e.g.*, precision farming, integrated pest management, improving water management practices, and soil conservation), it is possible to practice sustainable agriculture and allow farmers to cope with climate change. Science has revolutionized the way we farm and has led to significant improvements in food and nutrition security (more individuals have reliable access to nutritious food), as well as economic development (creating new jobs and increasing productivity).

ELECTRICITY

Electricity is one of the most important discoveries in the history of science, and it has transformed all aspects of life (*e.g.*, power to our houses, public places, transportation systems, and medical equipment to mention a few). Electricity is a form of energy that is generated from various sources (*e.g.*, fossil fuels, wind, solar, hydropower, and nuclear power), and currently power generation is being transitioned from antiquated methods (*e.g.*, burning coal and fossil fuels) to renewable sources. Electrical power is an area of major scientific research, which has led to many new technologies (*e.g.*, drawing water from oceans and rivers to supply to homes and other sources of renewable energy). Electricity is a fundamental aspect of modern life, which has tremendously shaped and been shaped by science.

TRANSPORTATION

Transport is necessary for the movement of people and goods from one corner of the globe to another. Science has led to significant progress in transportation technology in various aspects, such as efficiency, safety, environmental impact, and sustainability of transportation. For example, science helps to develop more efficient modes of transportation (*e.g.*, hybrid and electric vehicles), which reduce carbon emissions and transportation costs. Science also helps us understand the physical principles involved in transportation (*e.g.*, the laws of motion), leading to the production of vehicles with better safety features (*e.g.*, airbags and seat belts), resulting in systems that minimize accidents and injuries.

Finally, science also drives innovation in various advanced aspects of transportation, including (1) development of lightweight materials (*e.g.*, carbon fiber) that significantly reduce vehicle weight making them more fuel efficient; (2) design of modern transportation infrastructure (*e.g.*, roads, bridges, airports, and tunnels) and optimized construction material properties; (3) innovation of advanced modes of transportation (*e.g.*, airplanes, electric vehicles, high-speed trains, autonomous vehicles, and the hyperloop) to make transportation faster, safer, and more efficient. Overall, science is essential for transportation, and advances in science will continue to play a critical role in shaping the future of transportation.

SUMMARY

Although career scientists play an active role in the advancement of scientific knowledge, we all have a part to play in some way. As part of our daily lives, we apply scientific principles every day, whether we are conducting simple experiments, observing the world around us, or questioning the nature of the world we live in. As long as we cultivate scientific curiosity and promote scientific literacy, we will be able to promote a society that values evidence-based reasoning and embraces the scientific method.

Science plays a key role in advancing our understanding of the universe and natural phenomena, leading to improvement in our quality of life. In the most formal sense it is a systematic and evidence-based approach that allows us to make accurate predictions, develop new technologies, and solve complex global problems. Science is critically important to our society in countless ways, fundamental for human progress, and a crucial part of our modern society.

REFERENCES

Fähnrich, Birte, Clare Wilkinson, Emma Weitkamp, Laura Heintz, Andy Ridgway, and Elena Milani. "Rethinking Science Communication Education and Training: Towards a Competence Model for Science Communication." [In English]. *Frontiers in Communication* 6 (December 22, 2021).

Hoeg, Darren G., and John Lawrence Bencze. "Values Underpinning Stem Education in the USA: An Analysis of the Next Generation Science Standards." *Science Education* 101, no. 2 (2017): 278–301.

Jucan, Mihaela Sabina, and Cornel Nicolae Jucan. "The Power of Science Communication." *Procedia-Social and Behavioral Sciences* 149 (2014): 461–466.

Kaufmann, Alexander, and Franz Tödtling. "Science–Industry Interaction in the Process of Innovation: The Importance of Boundary-Crossing between Systems." *Research Policy* 30, no. 5 (May 1, 2001): 791–804.

Kouroussis, Denis, and Shahram Karimi. "Alternative Fuels in Transportation." *Bulletin of Science, Technology & Society* 26, no. 4 (2006): 346–355.

Linn, Marcia C., Libby Gerard, Camillia Matuk, and Kevin W. McElhaney. "Science Education: From Separation to Integration." *Review of Research in Education* 40, no. 1 (2016): 529–587.

Lipper, Leslie, Philip Thornton, Bruce M. Campbell, Tobias Baedeker, Ademola Braimoh, Martin Bwalya, Patrick Caron, *et al.* "Climate-Smart Agriculture for Food Security." *Nature Climate Change* 4, no. 12 (December 1, 2014): 1068–1072.

Marincola, Elizabeth. "Science Communication: Power of Community." [In English]. *Science* 342, no. 6163 (December 6, 2013): 1168–1169.

McFaddenĆ, Daniel. "The Behavioral Science of Transportation." *Transport Policy* 14 (2007): 269–274.

Peters, Glen P., Borgar Aamaas, Marianne T. Lund, Christian Solli, and Jan S. Fuglestvedt. "Alternative 'Global Warming' Metrics in Life Cycle Assessment: A Case Study with Existing Transportation Data." *Environmental Science & Technology* 45, no. 20 (October 15, 2011): 8633–8641.

Pittelkow, Cameron M., Xinqiang Liang, Bruce A. Linquist, Kees Jan van Groenigen, Juhwan Lee, Mark E. Lundy, Natasja van Gestel, *et al.* "Productivity Limits and Potentials of the Principles of Conservation Agriculture." *Nature* 517, no. 7534 (January 1, 2015): 365–368.

Rull, Valentí. "The Most Important Application of Science: As Scientists Have to Justify Research Funding with Potential Social Benefits, They May Well Add Education to the List." [In English]. *EMBO Reports* 15, no. 9 (September 2014): 919–922.

Spector, Jonathan M, Rosemary S Harrison, and Mark C Fishman. "Fundamental Science Behind Today's Important Medicines." [In English]. *Science Translational Medicine* 10, no. 438 (April 25, 2018): eaaq1787.

Tilman, David, Kenneth G. Cassman, Pamela A. Matson, Rosamond Naylor, and Stephen Polasky. "Agricultural Sustainability and Intensive Production Practices." *Nature* 418, no. 6898 (August 1, 2002): 671–677.

Tscharntke, Teja, Alexandra M. Klein, Andreas Kruess, Ingolf Steffen-Dewenter, and Carsten Thies. "Landscape Perspectives on Agricultural Intensification and Biodiversity–Ecosystem Service Management." *Ecology Letters* 8, no. 8 (2005): 857–874.

4 A Day in the Life of a Scientist

Samuel J. S. Rubin and Nir Qvit

The mindset of a scientist was discussed in general terms in the Introduction of this book. Briefly, the scientist must be driven to advance knowledge, however small or large in scale an individual piece may seem. The scientist uses a scientific method founded on rigorous skepticism and iteration to satisfy their curiosity and is not afraid of learning from failure. The scientist must be able to learn anything necessary to address a question of interest, and often must be confident enough to teach themselves. Regardless of their institutional setting, the scientist is obliged to communicate, teach, and pay forward knowledge gained through hard work.

WHEN AND WHERE?

Work hours of a scientist are variable, often independently set, and almost universally long, driven by one's own passion for their topic of study. Ultimately, one's hours are set by the nature of their science and the timelines necessitated to complete requisite experiments. For instance, a computational biologist or bioinformatician whose experimental medium may be largely accessible on a computer from any location at any time will be able to perform tasks with this flexibility regarding location and time, so long as they are making progress toward a defined goal and are subject to any particular institutional or supervisor policies that mandate work hours or location. Work using devices, chemicals, proteins, and other non-live reagents is probably the next most flexible, such that the scientist may perform work during typical scheduled hours. Use of cell culture requires regular intervention but can also usually be performed during regular scheduled hours. In contrast, scientists who utilize animal models in their experiments are often required to attend regularly to their animals for weeks to months before and during an experiment. This can range from daily to weekly responsibilities and may necessitate long or odd hours given that certain species have sleep/wake schedules opposite that of humans, although the yield of this work can be significant and impactful. Similarly, a researcher whose experiments depend on fresh human specimens is subject to the clinical availability of those valuable resources, and as such their work hours will be governed accordingly. Depending on the circumstances, clinical specimens may be available during predictable daytime business hours, or they could become available on an emergency basis. In the latter scenario, a researcher may find themselves "on-call" in case a precious

DOI: 10.1201/9781003301400-4

FIGURE 4.1   A typical laboratory bench, hood, and desk. Created with BioRender.com.

specimen becomes available. I recall many nights when I returned to the lab at 10 or 11 PM or even 1 AM to process a precious human specimen that I had been waiting for to perform a particular experiment. Given the spectrum of time commitment necessitated by different scientific media and the distinct capabilities of each method, a scientist almost universally utilizes more than one of the aforementioned techniques, often via a stepwise approach to maximize return on resources and time investment.

Laboratory research is often performed in a large building on an academic campus or in an industrial office park. These buildings usually contain multiple laboratories with varying degrees of interconnectedness ranging from open floor-plans to siloed lab bays. In our experience, open floorplans allow for the greatest amount of collaboration and happiness among researchers, although sometimes this is not possible due to availability of space or certain legal practices that prevent the sharing of information between teams on different projects. Small desks at the end of a large lab bench can be found in most lab bays. Sometimes there is also a separate office space with desks for computer work. Much time is spent at the desk – ideally adjacent to the lab bench and nearby other researchers, although sometimes in a separate office workspace or nowadays from home or another remote location – in order to accomplish many of the responsibilities and tasks outlined in the following section (Figure 4.1).

WHO, WHAT, AND WHY?

The best science is a team endeavor. Scientific teams generally follow a characteristic structure. The team leader is the principal investigator (PI), who is in charge of the laboratory. The PI could be a professor in an academic setting or a manager in an industry setting. Under the purview of the PI is a team of researchers with different levels and types of training, many but not all of whom are working in the lab as part of their ongoing training. Junior lab members may in some cases include high school or college student interns, but more often start with bachelor's degree-trained research assistants (RAs) or technicians (techs). In some cases, these RAs or techs may have the longest tenure in the particular lab and thus may not be so junior in that sense.

Next are graduate students (grad students), including master, MD, PhD, and other graduate degree-seeking trainees. Following grad students are postdoctoral fellows (postdocs), another level of advanced training after a doctorate degree program. Grad students and postdocs are often the powerhouses of the lab in terms of their ability and motivation to make contributions. Senior scientists are longer-term employees that are also very productive given their advanced degree and postdoc training often in the same or adjacent lab. The lab manager may be one of the aforementioned individuals or may be a separate position, which is integral to the function of the lab. The lab manager is generally responsible for performing or coordinating other lab members to perform: lab resources inventory, reagent stock, supplies ordering, lab cleaning, equipment maintenance, etc. Some labs have only RAs and postdocs, while others have only grad students. There are different team styles across institutions and fields, but lab members generally fall into the aforementioned categories with characteristic training and responsibilities. Although there is some inherent hierarchy based on level of training, in science there is frequently an understanding that most lab members bring unique expertise they can share and teach other lab members across level of training. In this sense, lab members are colleagues across levels of training, and in some cases even quasi-peers.

Time spent as a scientist is dedicated to a wide array of tasks. Review of the primary literature is critical for understanding your field, identifying what has already been done so that you do not find yourself "reinventing the wheel," and determine what has yet to be discovered. It is also critical to stay up to date with the literature in order to remain current with cutting edge technology and findings. Scientists find themselves reviewing primary literature across new niches throughout their careers, since science is an inherently iterative process that often takes investigators in new or unanticipated directions based on experimental findings, as discussed in later chapters. These days, literature review is accomplished almost universally via the Internet, and investigators do this from the comfort of various locations.

Seeking and applying for funding for resources to perform experiments that address a particular question and hypothesis is critical for the practice of academic research. In industry, it is also often important for an investigator to be able to justify spending of funds on a particular project or experiment. Funding is often a somewhat circular process in that one must combine knowledge of the literature with preliminary or pilot data (which requires funding to obtain) into a compelling presentation in order to convince a funding body (government institute, private donor, foundation, non-profit, or other entity) to provide a pre-specified amount of support via financial and/or tangible resources for a particular amount of time. Thus, communication in writing is crucial for a scientist to obtain the necessary resources to advance knowledge in their field.

Designing a study and planning experiments based on a question and hypothesis is also usually accomplished on the computer in this age, although it can be tremendously helpful to be in a lab setting during this process in order to discuss with colleagues and reference various reagent or equipment parameters found in the lab area. Conducting experiments, unless purely computational, necessitates a location based on the resources involved. The majority of one's experiments will ideally take place in their lab at a bench or other structure, for the same reasons including proximity to colleagues and resources for brainstorming and trouble-shooting. Certain experiments must be performed in separate cell or tissue culture rooms, core facilities with specialized equipment, or even at secondary locations with particular resources or expertise. The end product of the experiment is raw data.

Analyzing and interpreting data is generally performed via computer, unless highly specialized equipment is required due to the nature of the raw data. Some scientists choose to analyze data remotely due to the convenience of choosing the time and location, but it can be tremendously helpful to perform this part of the study in the proximity of colleagues, especially for the more junior researcher. The best interpretation of data, even for advanced investigators, includes discussion with colleagues.

Findings are summarized and presented verbally and graphically via writing reports, including both internal and external publications depending on the nature of the work and the institution (*i.e.*, industry vs. academia). Dissemination of knowledge is an important part of the scientist's job and can include written publications, as well as informal meetings and talks, poster presentations, and formal talks or lectures – at one's own institution, another institution, and/or conferences. In addition to sharing one's own findings, conferences can provide opportunities for travel, networking with others in the field or adjacent fields, exposure to published and unpublished findings, meeting potential collaborators and mentors, identifying prospective job opportunities for oneself, and connecting with potential new hires or future trainees for one's own program.

Teaching is another important responsibility of the scientist in order to disseminate knowledge and pay it forward, as previously discussed. The traditional classroom – at various levels – is one setting where teaching takes place. Another setting where teaching takes place on a regular basis is the lab. During the course of training, individuals gain unique exposures to specific areas, which they can subsequently share with others in a lab group. Thus, lab members of all levels teach colleagues in the lab various things, such as scientific mechanisms, skills, processes, protocols, and equipment operation. The same occurs in both academic and industry lab settings, as scientific investigation is inherently a team effort in the lab group.

IS SCIENCE FOR YOU?

The scientist wears multiple hats, plays many roles, takes on responsibilities that require skills ranging from analytical critical thinking to public speaking and social networking. This expertise can be gained formally or informally, and you must be willing to learn almost anything – that is the fundamental tenet of being a scientist.

5 The Scientist's Skillset

Samuel J. S. Rubin and Nir Qvit

GENERAL COMPETENCIES

Several general competencies closely related to our core philosophies for the scientist of curiosity, skepticism, and learning are particularly helpful when considering skills of the successful scientist. While curiosity and skepticism are central to the essence of science, these practices are also integral to the conduct of science. Without inherent curiosity and skepticism, it would be very challenging to identify new questions and to continually verify, update, and correct knowledge. One manifestation of this philosophy is the power of limiting one's assumptions and instead asking more questions. This is akin to a child making an observation and asking, "Why?" If we are presented with information and ask this question, then we can often identify opportunities for investigation. Sometimes the information presented will not be as certain as one might like to believe if the question, "Why?" cannot be answered, and sometimes the process of asking this question will bring about adjacent questions which remain uncertain. This is not to say that information is uncertain if there is no readily available straightforward explanation – sometimes getting to the bottom of a question requires extensive reference material and/or consulting an expert. But in general, taking the time to ask, "Why?" before taking a piece of information at face value will save you time and energy.

Being a lifelong learner is also key to the practice of good science. Because new data often takes the investigator in different directions – familiar or unfamiliar – you must be willing to seek your own resources and use those to teach yourself a new area. You need to find your own tools to become an expert in something new to you, quickly. This does not mean that you should not seek help; in fact, you should seek assistance from other experts and teachers – this is part of the process of seeking your own resources. Another resource you must be willing to utilize is "on the job learning." You must be willing to dive in and learn as you do something new, gaining knowledge and insight from a variety of resources as well as trial and error. Perhaps most importantly, it is particularly helpful if you can learn to learn from other people's mistakes, seeing things done wrong by others, or simply observing someone else's practice and improving upon that practice for your own approach. This process will save you immense time and headache, and you will be a much more efficient investigator for it.

In order to accomplish the aforementioned practices, you must have forethought, you must take initiative, and you must be creative. Forethought can be a simple as recognizing that you need to keep record of all volumes used in an experiment for

DOI: 10.1201/9781003301400-5

calculations later. As you gain more experience, forethought will include planning the experiment after the next experiment while you are still just beginning the first experiment. Taking initiative is key for acting upon these ideas, and you have to be fearless. It is important not to let fearless initiative turn into recklessness, but we will save that for a later discussion. Creativity will allow you to solve unsolved problems, create new methods that better accomplish a common goal, and apply research findings to address unmet needs across fields.

TECHNICAL EXPERTISE

It is impossible to encompass the entire breadth and depth of technical expertise that a scientist will curate during their career (not to mention across scientific disciplines), and much of that relates to continued learning and adaptation. However, several common themes are helpful to consider across scientific disciplines. Analytical and critical thinking are central to many tasks, as is attention to detail. Precision and meticulous record keeping are imperative for interpretability and reproducibility. Whether motivated by discovery and/or application, research efforts benefit from these core practices. No single technical skill is required to break into the world of science. The skillset will come with time and evolve over the course of one's career. But one must be motivated and equipped with the capacity to build and maintain a toolbox of detailed, analytical techniques.

Being facile with computational methods greatly increases one's value and capacity in today's climate of big data. With advances in technology, we as a field are now collecting exponentially larger datasets. Thus, analytical techniques must adapt and grow in order to meet the input of big data. A basic understanding of statistical methods and when to apply different approaches is foundational to the appropriate analysis of large datasets, and too often this fundamental point is overlooked. Being able to understand sufficient code in order to analyze data in R, MatLab, Python, or another program will expand your capability for scale and your span across fields. A deeper understanding of computer science and machine learning will drastically expand your horizons and allow you to utilize and compete with cutting-edge techniques. We cannot stress enough how transformative a fundamental understanding and competency in computer science empowers the young scientist for success, and we will return to this point in further chapters.

Communication skills are also important for the successful funding and conduct of research, replication, and dissemination of findings. In order to discuss ideas with colleagues, share results, and disseminate findings one must be facile at communicating in writing and orally. Explaining one's science is critical for writing grant applications, scientific papers, and other reports. Giving oral presentations is also an important part of networking in the arena. Without clear documentation of an experiment, results may not be replicable and thus devalued. Being able to convey complex scientific concepts to experts in your field as well as lay audiences, using different vocabulary for each context, will set you apart when applying for funding, publicizing results, and applying findings to real-world problems. Too often a brilliant scientist overlooks the significance of good communication across audiences of different training and technical expertise.

HELPFUL EDUCATIONAL AND CAREER PRACTICES

Like communication, management skills are important for success in the team-based environment where science is conducted. There are entire books dedicated to good management practices, and we do not attempt to provide a comprehensive guide to this realm. However, it is worth emphasizing that management deserves special attention and thought from scientific investigators. Managing down describes how an individual deals with subordinates, which might be the type of management most frequently associated with this term. A growing appreciation for managing up has characterized recent developments in the field. Managing up generally involves taking an active rather than passive role in working with superiors. In addition to good managerial promotion of working conditions, workplace culture, time management, recognition of contributions, active leadership, setting and adjusting expectations, regular feedback, and acceptance of limitations, managing up requires one to understand the specific interests of superiors and balance this with self-advocacy and asserting one's own needs in order to be successful in the workplace. We also feel that it is important to consider managing left and right; science is a collaborative endeavor, and team management of colleagues across the spectrum of roles is critical for success and enjoyment. In essence, taking an active role as appropriate to manage the jobs of those superior, subordinate, and adjacent to you will make your job easier. This process relies upon clear roles and responsibilities, as well as prompt, clear, and compassionate communication about the work. Feedback is an essential part of this process, both structured and unstructured. When addressing feedback, it is important to ask others to share their perspectives, and to solicit feedback whenever offering feedback. Taking responsibility is one of the most powerful ways to address feedback broadly.

Taking time to celebrate accomplishments is another helpful practice to find joy in the scientific process. Research can be circuitous, seemingly endless, and frustrating at times. Science is also tremendously rewarding. It is important to celebrate milestones and accomplishments, small or large, especially as the large ones do not come frequently. Celebrate hard work, and also take advantage of times to relax and rejoice from work. Utilize time off to travel or dedicate time to pleasures other than work and to individuals significant in your life. While the "work hard, play hard" motto is trite and overused, it is important to find balance in your life. Time away from the intensities of work or studies can often provide time to reflect and gain new perspectives that improve your ability to perform well at work or in training.

COPING SKILLS

While there are many "ups" in the career of the scientist, there are inevitably "downs" even for the most successful. To traverse these obstacles in stride, one must take active measures to promote their well-being and avoid burnout or address it early. As our society promotes the expansion of workaholic culture, it is important to be mindful about work-life balance. The WHO defines occupational burnout as a syndrome conceptualized as resulting from chronic workplace stress that has not been successfully managed ... characterized by ... feelings of energy depletion or exhaustion; increased mental distance from one's job, or feelings of negativism or cynicism related to one's job; and reduced professional efficacy. (Burn-out an "Occupational Phenomenon": International Classification of Diseases, 2019)

Unfortunately, occupational burnout is an increasingly common phenomenon. The signs can be subtle and not always recognized as related to burnout, including irritability, cynicism, feeling unmotivated, low energy, difficulty concentrating, lack of satisfaction, disillusionment, change in sleep habits, adoption of unhealthy coping mechanisms like substance use, and even organic symptoms like headache or stomachache. In this case, a "work hard, play hard" mentality can backfire and lead to a cycle that further perpetuates burnout. We do not claim to have cracked the code on burnout or life happiness, but from our experience we have found success with an algorithm involving a balance between daily time spent on: intellectual activity, physical exercise, nutrition, sleep, and play. The specific distribution across these factors will differ by individual and by life stage. The key is to find time to enjoy the journey of science, of one's career, and of life. Too easy is it to lose track of the journey and focus entirely on the destination, or goal – be it promotion, discovery, or other general motivation. It is important to have this motivation and an overarching long-term goal, and it is equally important to see the forest for the trees. Enjoy the journey. Take pride in daily achievements. Delay some gratification, but not completely.

SUMMARY

The scientist's skillset is a continuously evolving toolbox. The scientist must be willing to become a jack of many trades. Many of these skills are not unique to the work of a scientist, but their application and their continual adaptation is a special challenge of the scientist's work. A wide variety of technical skills, as well as communication and managerial competencies, are key for success as a scientist. None of us start out perfectly. Much is gained by trial and error. Being thoughtful in that process is helpful, and we hope that we can save you some of the hassle by illustrating key lessons in advance. Learning other lessons on the job is unavoidable, if not expected, as it is a fundamental necessity to be open to failure and use lessons learned for growth. Stanford psychologist Carol Dweck and colleagues coined the term "growth mindset," which is important for adapting and improving oneself along this journey to success we call life. No more legitimate for any field is this than for science. Not only is our work a continual process of experimentation and iterative improvement, but often so too is the scientist's career path.

REFERENCE

Burn-out an "Occupational Phenomenon": International Classification of Diseases. (2019). World Health Organization. https://www.who.int/news/item/28-05-2019-burn-out-an-occupational-phenomenon-international-classification-of-diseases

6 The Conception of a Career in Science

Nadav Qvit

EARLY SCHOOL YEARS

The first few years of grade school are great for introducing children to a diverse range of subjects. Lessons in schools are complemented by children's natural curiosity, and as such the most effective lessons for these ages include those that encourage students to try things out for themselves.

One of a school's main goals should be to confer inspiration to its students, whether it be in literature, sport, volunteer work, and/or science. Experiential learning induces more interest and better understanding of the lesson subject. Because of increased understanding and appreciation for the importance of experiential learning, more and more schools today are using this teaching style.

In addition, science fairs and extracurricular opportunities allow young children exclusive access to advanced concepts and knowledge and often support near peer mentoring. Participating in scientific activities outside of school cultivates children's curiosity as well. These venues can include exhibitions and fairs, which are often presented in libraries, colleges, and other communal areas that are devoted to education. Museum visits are a wonderful opportunity for children to interact with information and items that are less commonplace, and also serve as a place to meet with various guides and experts on specific subjects.

My most distinct memories from science classes at school are those in which we conducted our own experiments or were able to interact with the study material in a hands-on manner. I personally am able to understand what I am studying more efficiently when doing experiments myself, rather than reading about someone else conducting experiments. Two instances stand out in my memory. One of them includes an activity I participated in during 3rd grade science class in which we were supplied with kits containing owl droppings and we were given the opportunity to analyze what the owl had eaten according to the bones we found. The other experience was an activity in which I extracted strawberry DNA in a plastic bag during a fair organized by Stanford University. Even today, my understanding of subjects I read about in textbooks and articles is substantially supplemented by experiments I conduct.

LATE SCHOOL YEARS

Elementary schools, as well as most middle schools, offer compulsory "science" as one of the classes that students must complete. These classes generally cover basics

DOI: 10.1201/9781003301400-6

in biology, chemistry, and physics, without diving in-depth into any of these topics. In high school, in contrast to the generality of "science," students have the option to focus on specific fields for deeper learning in later years. These classes offer students who are interested in specific topics opportunities to learn more about them at a higher level.

STANDARD LEVEL HIGH SCHOOL COURSES

A diversity of high school education systems worldwide allows students to receive an education relevant to their cultural, geographical, and political environment while acquiring knowledge in universal subjects such as mathematics and science. In the United States, standard high school mathematics typically includes classes in trigonometry, pre-calculus, and in some cases calculus, while science usually includes classes such as introductory biology, chemistry, and physics. There are additional opportunities at some schools for students to take more advanced courses in these subjects as well as adjacent topics in science, technology, engineer, and mathematics (STEM) fields to expand their knowledge. Class options differ significantly based on type of school in the United States, which is often a motivating factor for choice of school among students interested in particular fields who are able to choose between multiple different schools (although sometimes there is only one option). Other countries have different systems of allocating students to schools and classes, which vary extensively across the globe.

THE MATRICULATION EXAMS IN ISRAEL

The matriculation program is offered in high schools nationwide in Israel to allow standardization of grading. Students take classes in seven mandatory subjects, and at least one personally selected subject. Out of the seven mandatory subjects, five of them – history, literature, language arts, citizenship, and religious studies – are considered core humanitarian subjects. The other two subjects are math and English and are separated into levels (3, 4, and 5 points) in order to ensure that everyone can study them at an appropriate capacity. The personally selected subject(s) can include scientific subjects such as biology and chemistry, artistic subjects such as theater and photography, as well as many others. There is no such matriculation examination system in the United States, although some other countries employ an analogous process.

ADVANCED HIGH SCHOOL COURSES

In addition to specified classes across a menu of topics, some high schools offer advanced classes, which allow students to delve deeper into the subject matter, learn more, challenge themselves, and demonstrate their competencies via standardized exams. Several paradigms are presented here for advanced high school courses; however, it is important to note that some countries and regions have different versions of these programs.

In essence, these programs allow high school students the chance to study an advanced curriculum, and students may use their participation and grade in these curriculums to demonstrate proficiency in a particular field, apply to subsequent studies (training programs, universities, etc.), and at some institutions (depending on their policies) they may be awarded university credits. Thus, enrollment in advanced courses and completion of exams is a means, where available, to both gain and demonstrate advanced level proficiency in certain subjects during high school studies.

International Baccalaureate (IB) Courses, Exams, and Programs

The International Baccalaureate (IB) was originally borne out of a program created to standardize courses and assessments for high school level students across nationalities. Currently, the IB is based in Geneva, Switzerland, and includes the IB Primary Years Programme for students ages 3 to 12, the IB Middle Years Programme for students ages 11 to 16, and the IB Diploma Programme for students ages 15 to 19. The IB Diploma Programme offers advanced courses and standardized exams to students across a comprehensive array of subjects. What sets the IB curriculum apart are the comprehensive programs administered by the organization. The IB Diploma Programme maintains a rigorous curriculum that requires students to study and pass examinations in six subject groups, including studies in language and literature, language acquisition, individuals and societies, experimental sciences, mathematics, and the arts. Diplomates must also complete three core requirements, including creativity, activity, service (CAS) projects involving community service, an extended essay (EE) akin to a thesis, and theory of knowledge (TOK) studies on fundamentals of philosophy. Overall, the IB Diploma Programme offers a comprehensive, advanced curriculum that instills well-rounded and critical thinking skills in students seeking rigorous inquiry. Completion of IB exams and earning the diploma affords students university level credit at some institutions and also provides students with a standardized qualification that allows them to apply for subsequent training programs across the globe, in many instances without taking specific national entrance examinations.

Advanced Placement (AP) Courses and Exams

The Advanced Placement (AP) program is based largely in the United States and Canada. Generally, the AP curriculum includes specialized courses offered at certain high schools that teach an advanced curriculum in a particular subject, from which there are a variety of options across fields of study – dozens ranging from arts, languages, and history to math and sciences. Most students who elect to enroll in AP courses will do so as an alternative to the standard level course for a given subject, although some will take the AP version of a subject following completion of the corresponding standard version.

Cambridge Assessment International Education (CAIE, formerly CIE)

The CAIE is a UK-based organization and department at Cambridge University that provides educational programs – including lessons, examinations, and qualifications – in an international setting. Each program is tailored for certain years in a student's education. Programs are designed for use from age 3 up to high school graduation. These programs offer a broad and balanced curriculum at early stages and subsequently allow students to choose what to study from a wide choice of subjects. One aspect that distinguishes CAIE from other programs is their flexibility and that they allow international education systems to implement their own culture into curricula. Successfully completing the CAIE program provides graduates internationally recognized qualification – and in turn the opportunity to continue studies at a university.

EARLY PARTICIPATION IN COLLEGE CLASSES

A common course of action for students wanting to learn more advanced material is to enroll in college classes before graduating high school – in the United States, this is called dual-enrollment or concurrent enrollment (NACEP, n.d.). This allows students to access advanced material while gaining college credits before officially beginning college, and some may even transition directly into a bachelor's degree program with partial credit as soon as high school ends.

These college courses can be instead of, in addition to, or in continuation of advanced high school level classes. Both arrangements have their advantages and disadvantages, and it is important to know them before deciding which of them to enroll in, and when. There is great diversity in programs, and when deciding upon each of them, key factors which should be considered include academic level ("Is this program too difficult/easy for me?"), peers and colleagues ("How large are the classes?", "Are my classmates the same age as me?"), intellectual benefits ("What will I learn?"), and academic advancement benefits ("Will this make my CV stand out?").

Taking college courses throughout high school can open up a plethora of opportunities later on. First, this makes a college application, job CV, or other appeal more impressive, putting the applicant ahead and enhancing their chance of being accepted. In particular, the college in which the classes were taken may be more inclined to take on someone who has already studied there. In addition, participating in a course allows students to see what is needed for them to succeed in college and may also help them choose what subject they want to major in and what institution they wish to join.

Throughout high school, I participated in a program named Odyssey, which allowed me to enroll in college courses along with peers from my grade. This allowed me to study the subjects that I loved at a level that was higher than I could have received elsewhere, and I was able to do it with friends that were my own age and are now some of my closest confidants. In addition, the college credits that I earned through this program granted me the opportunity to graduate with a bachelor's degree only one year after finishing high school.

ACADEMIC COMPETITIONS

Academic competitions are a great way for students to test their knowledge, as well as compete with other students and build camaraderie. These competitions include school science fairs, regional competitions, and international-level Olympiads.

Competitions in a variety of topics are hosted worldwide, involving both the science and the humanities. Other than the acclaim that can be earned in these events, they are a great way to meet other people who share common interests.

Success in competitions requires a different set of skills than is needed for success in school and college. Competitions require active problem solving, thinking out of the box, and extrapolating knowledge quickly, instead of memorization. In order to prepare for a competition, you can train by solving questions from past years, as this will allow you to figure out how best to answer possible questions. Another way to prepare for a competition is to study with colleagues, whether as part of a self-assembled study group or in official sessions organized by the competition when available. This manner of studying can be especially beneficial for competitions, a setting where comprehensive feedback is not always given.

Some competitions have groups, rather than individuals, compete with one another. This is often seen in technology-oriented competitions such as First Tech Challenge (https://www.firstinspires.org/robotics/ftc), as their assignments require a diverse set of technical knowledge and skills.

I highly enjoyed competing because it allowed me to measure my problem-solving skills using knowledge in the subject matter, in addition to making friends with similar goals and interests to mine. In fact, some of my closest friendships are with people I've met while training and competing. I completed competitions in math and chemistry, as well as biology, where I represented Israel twice at the International Biology Olympiad, coming back from both with a silver medal. The experiences I earned while training and competing have been incredible and will undoubtedly stay with me for the rest of my life.

EXTRACURRICULAR ACADEMIC PROGRAMS

Special programs that aim to enrich students and support their curiosity have been developed for a wide range of topics. Emphasis has been placed on subjects ranging from the sciences to humanities and leadership skills, for the purpose of education, competition, creation, or a combination of the three. These programs allow participants to gain skills, awards, entertainment, and friends. To get the most out of it, participants must be willing to dedicate time and energy to their program of choice.

As a consequence of the vast selection of programs, the advantages each one bestows upon its participants must be considered when applying. Reading on the program website, speaking with a program manager or graduate, etc. can help reveal more features and nuances to prospective applicants.

For many programs, resources are limited, causing them to constrain the number of participants. This is often done through entrance testing to ensure that the participants chosen are those that best fit the program. To successfully pass these tests, applicants should be familiar with the testing content, stages, and strategy. Training for tests can help prepare applicants, as strategies can be different depending on what type of test is administered.

Participating in formalized programs can offer a variety of benefits that last beyond the programs themselves. Skills and knowledge developed during the experience add to a more comprehensive base for future work. Some programs show how science and technology are intertwined, and the understanding of this principle can allow a smooth transition from academic study to applied technologies. The prestige of some programs shows examiners that participants have additional merit. Belonging to a group of program graduates also fosters connection and networking with others who have participated in the same or similar programs.

I have participated in two major extracurricular academic programs. The first was called Odyssey, where I took college classes throughout my high school years. The second was called the Biology Olympiad, where I studied and competed in both the Israel National Biology Olympiad and the International Biology Olympiad. Whereas Odyssey was purely educational, the Olympiad had an extra element of competitiveness. This was evident in several places, including the examination process. The Olympiad entrance tests were purely based on knowledge of the subject, while Odyssey called for social skills as well. Both experiences taught me a great deal and enhanced my knowledge on a variety of subjects, in addition to training me in unique skills and techniques that I will no doubt use for years to come. In addition, Odyssey and the Olympiad allowed me to join a nationwide network that connects all graduates of similar programs, through which I can keep in touch with friends I made in these programs as well as meet new people with whom I can collaborate and/or befriend. For other programs and types of experience, see Chapter 19.

THE INTERNET AND DIGITAL RESOURCES

The advancement which has proven most pivotal in allowing youth to engage in academia has been the invention and propagation of the Internet, especially the easy access to information that it grants. The advent of the Internet afforded multiple advantages. In particular, the ease which accompanies proliferation and facilitation of study materials and additional methods available via the Internet have transformed when, how, and what young students can learn.

There are many platforms that have taken advantage of this progress in accessibility, having been built with the intent of educating young children. Several of these resources include Khan Academy (https://www.khanacademy.org) and Brilliant (https://brilliant.org), as well as a handful of content creators who use YouTube, such as Hank and John Green who host several channels including the Crash Course channel (https://www.youtube.com/@crashcourse), Dianna Cower from Physics Girl (https://www.youtube.com/@physicsgirl), Joe Hanson from Be Smart (https://www.youtube.com/@besmart), and various others.

These electronic resources have helped me understand the world around us and have also instilled in me a sense of passion, awe, and curiosity. Khan Academy has given me a rigorous understanding of math and physics with their curriculum. Their clear explanations and vast collection of problems allowed me to learn and practice the subjects that I enjoyed at the pace that fit me. Meanwhile, the other videos I spent hours watching, while not always being rigorous in their explanation, contributed to my desire to study and learn and introduced me to additional fields and disciplines.

STUDY TECHNIQUES

Just as each person has differing hobbies and passions, the techniques we use to study vary. In order to study efficiently, you must figure out which method(s) of studying works best for you, which is best explored early. To do this, parameters including study method, location, session length, and study partners must be considered.

Along with the digital age came an increasing variety of methods by which to learn. Massive Open Online Courses (MOOC) and Open Courseware (OCW), Online Classes (*e.g.*, through Zoom), interactive lessons (*e.g.*, Brilliant) and many others have all joined the classical classroom and textbook as viable learning modes (Reich, 2015). In addition, there are differences within each of these methods that depend upon the instructor, as well as the medium through which the lesson is held. In light of these and other approaches, you can sample and then choose the method by which you learn best, based on its advantages and disadvantages unique to your preferences.

Different people study best under distinct conditions. In order to figure out in what conditions you study best, it is useful to try out several environments – varying them until you can find those best fit for you. Several things must be considered: First, the place where you study – at home, in a library, at school, etc. Second, study partners – friends, classmates, a tutor, or alone. Third, session length – how long do you study at a time, how long are the breaks, etc. For example, I get distracted when studying with friends, and as such prefer to study alone. For a similar reason, I can focus better when studying at a library. Moreover, I prefer to set my study session by a task – such as reading a chapter or solving a certain number of problems – rather than based on time. Each individual must figure out what conditions are most beneficial for their studies.

Studying a subject most effectively often calls for a comprehensive, standard, recognized source or sources. One of the advantages of schools and colleges is that these institutions readily provide teachers, textbooks, and other sources, allowing students easy access to information. However, when studying a subject alone these sources are not necessarily on hand. Thus, being able to find these sources without the assistance of school is crucial. There are several options. The most trustworthy is to speak with someone who knows the subject – a teacher, professor, mentor, etc. Another option is to check a library; librarians often know which books may be the best fit to learn a subject. Of course, the Internet – forum pages, blog posts, journal articles, social media, and many more places – may also be used as a source. It is usually most beneficial to gather information from several sources, as any one of them may have false or limited information (this is especially important when using online sources).

CONCLUSION

This chapter traversed the path of young students interested in the sciences and discussed several programs in which they can participate and paths they might take. During the first several sections of this chapter we considered how students learn throughout school – from feeding the curiosity of elementary school students up to taking advanced courses in high school. Next, we presented opportunities to further study outside of the typical school curriculum, such as special programs, competitions, and early college courses. Last, we reviewed additional aspects of learning, including how to work with digital resources and navigate self-study. It is important to maintain perspective early on during the beginning of a young scientist's career and journey. There are a limitless number of ways to continue into more advanced studies and pursue study or research in the sciences. Hopefully, this chapter gave some insight as to how the earliest stages of a career in science can manifest and how to best utilize and enjoy these periods.

REFERENCES

NACEP. (n.d.). *What Is Concurrent Enrollment*. National Alliance of Concurrent Enrollment Partnerships. Retrieved September 8, 2023, from https://www.nacep.org/about-nacep/what-is-concurrent-enrollment/

Reich, Justin. (2015). Rebooting MOOC Research. *Science*, *347*(6217), 34–35. 10.1126/science.1261627

7 Finding the Right Training Program

Samuel J. S. Rubin and Nir Qvit

NETWORKING

One of the first steps in finding a new opportunity, whether training program or another position, is to network. Expanding and consulting your network of peers, colleagues, formal and informal mentors, and all kinds of individuals in adjacent or peripherally related positions is incredibly valuable. Many opportunities you seek are only learned by word of mouth, and for this reason networking is critical. Even if an opportunity is readily available public knowledge, you can learn about the nuances of what may be involved with applying to, obtaining, and maintaining such a position. In addition, personal network contacts can be tremendously useful when competing for desirable positions. Insider information, an informal endorsement or "good word" on your behalf, or simply an introduction on your behalf can make the difference when coming from someone "in the know."

Some of the best places for networking are career fairs, academic conferences, specific talks, and industry expos. However, networking can also occur spontaneously if you happen to meet a relevant individual in another setting, and frequently you won't realize how relevant they are until you start talking and find out his friend's sister does exactly what you want to do, and "oh, by the way, can you join her and me for coffee tomorrow?" Talk to as many people as possible. Ask questions and be curious. Meet people. Generally, they want to help you and they want to make connections and network too. Although it's human nature for many people you meet to want to help you, it's rare for someone to do so completely purely without some benefit to them. So, see what you can offer in return, even if that simply involves gratitude to make them feel good, and keep in mind that if someone is especially persistent with a seemingly selfless offer, then there might be more to it. Most of the time, individuals can help you and you can help them in a relatively balanced manner. Even if you feel they are really going out on a limb to help you, you're also helping them because it reflects well on them to be seen helping others and to present promising candidates. Take people up on their offers, within reason, as you never know where the road will take you. Try to avoid promising anything specific in return, unless you are certain that you can deliver on the promise. It's always better to under promise and over deliver. You can always use the rationale of your curiosity and learning in order to request help with something so that you don't commit yourself to necessarily going down the path if you learn more and decide that it's not for you.

DOI: 10.1201/9781003301400-7

When preparing for networking, it can be helpful to carry some business cards with your name, contact information, and general field or position. It never hurts to have a few in your wallet too because you never know when you might encounter someone you'd like to connect with. For more planned networking settings, have a curriculum vitae (CV) (see the section "Curriculum Vitae" below) ready to send as a follow-up email or even printed to hand a key contact. Print it on thicker, brighter paper to make an impression, and try to avoid double-sided printing.

During your encounters with individuals, less is generally more. Hold your proverbial cards close to your chest. That is, be friendly and engaging and discuss your interests or current work in broad terms, but don't over share. Just like you never know who will turn out to be the most helpful to you, you also never know who may be a competitor or other entity that you'd rather know less about you. That doesn't mean you can't be cordial or even friendly. But do so without over sharing, at least until you are confident that an individual is trustworthy and honest. Never say anything that you wouldn't want anyone else to hear, since you never know who your audience will tell or who might be able to overhear you. In science, kindness and curiosity generally get you the farthest.

IDENTIFYING AND CHOOSING SPECIFIC TRAINING PROGRAMS

There are multiple approaches to finding training programs, and you should consider utilizing all of them in parallel. There is also significant overlap between the processes of identifying training programs and choosing one to join if/when you receive acceptance(s). Spend time doing research online to identify training programs, talk with mentors, talk with colleagues, and ask network contacts for their ideas and recommendations. Some of the most important sources of information about specific training programs are current and former trainees, so ask individuals in positions ahead of you where they trained and why, what they liked and did not like, and where else they considered training and why they did not end up in those programs.

The "best" training program will be different for each individual. The best option is the one with the right fit, where you will be the most successful, be the happiest, and have the resources you need. This fit will depend on individual values, priorities, and interests, which vary. Some important elements we recommend considering are listed below for reference. These should be re-ordered based on individual priorities and needs, and we expect this will result in differences between individuals. For convenience, we separated factors into academic, career, and creature comfort categories in order to help frame your consideration. Some elements may seem negligible to one yet extremely consequential to another, and neither are incorrect. It's important to prioritize what matters to you most, since that will best set you up for success.

- Academic:
  - Rigor of the training (*e.g.*, depth, breadth, etc.)
  - Research resources (*e.g.*, laboratories, instruments, core facilities, vivarium, statistical support, information technology, etc.)

- Ancillary resources at the institution (*e.g.*, workshops, library, experts in different fields, etc.)
- Courses (*e.g.*, availability, class size, grading systems, requirements, electives, etc.)
- Teaching (*e.g.*, requirements, opportunities, settings, etc.)
- Grant funding available to the institution and trainees
- Scholarships available to trainees
- Accessibility of excellent mentors in your field
- Opportunity for graduated autonomy
- Career:
  - Resources and support for career development
  - Track record of successful graduates (who have gone on to work that aligns with your goals)
  - Reputation of the institution (*e.g.*, rankings, notoriety, etc.)
- Creature comforts:
  - Familiarity with the institution
  - Work-life balance
  - Geographical location
  - Cost of living
  - Cost of tuition (if applicable)
  - Salary (if applicable)
  - Benefits (if applicable)
  - Accessibility of other resources important to your lifestyle (*e.g.*, restaurants, parks, outdoor activities, bodies of water, etc.)
  - Campus atmosphere (the physical spaces where you will spend the majority of your time)
  - Community environment
  - Proximity to friends
  - Proximity to family
  - Preference of partner(s) (if applicable)
  - Religious life (if applicable)

APPLYING TO TRAINING PROGRAMS

General best practices for the application process to any type of training program include planning well in advance for fulfilling the requirements. Regardless of rolling admissions or other policies, it is usually beneficial for the applicant to submit their application materials as early as possible, without compromising the quality. If possible and appropriate, it can also be helpful to personally contact individuals associated with the training program or institution, such as a professor or investigator. This can be done if you meet them at a conference or lecture, or you can "cold email" them due to shared interest which you could discuss with them. Either way, let them know you are applying, why you are excited about the prospect of joining the training program, and attach a copy of your CV (see the section "Curriculum Vitae" below). Also consider reaching out to such contacts before submitting application materials to inquire about any suggestions or advice they may have, and then follow up again once you have submitted your application materials. Most training programs require some or all of the following application materials, including letters of recommendation (LORs), personal statement, research statement, CV, and additional information. Be especially kind to the coordinators and assistants you interact with through the application process; they can put your application on the bottom of the pile or the top of the pile.

LETTERS OF RECOMMENDATION

Almost all training programs require at least one if not multiple letters of recommendation. It is generally considerate to solicit LORs at least one month, if not more, in advance of the deadline. It is also best practice to follow up with friendly reminders to the letter writer as the deadline approaches. When considering who to ask to write a LOR, it is important to think about and ask who would be willing to write a *strong* letter of recommendation on your behalf. A lukewarm letter will do more harm than good. Also consider what reputation a potential writer has in the community or field of the training program, as this will affect their impact. In addition, try to submit as many LORs as allowed, since submitting less (even if some are optional) can be interpreted as not having enough individuals who can write a *strong* LOR on your behalf. Consider a mentor or more senior individual with whom you have studied or worked who can speak to your qualifications and potential. After a letter writer agrees to write a *strong* LOR on your behalf, it is prudent to send a copy of your CV (see the section "Curriculum Vitae" below) and any draft application statements (see below) so the letter writer can align their LOR with your application rationale and so that they can also give you feedback on how to improve your draft statements. Some letter writers will request that you compose a draft letter for them to review. Try not to take offense, as this may simply indicate their level of time commitment and/or common practice in their field. It can be challenging to write a glowing letter about yourself, especially if you need to do this for multiple letter writers, so ask for a sample letter that they wrote for another individual so that you can understand and emulate their voice in your letter. Always waive the right to view the final submitted letter, as the letter will have significantly less meaning if you do not waive this right.

PERSONAL STATEMENT

The personal statement essay is an opportunity to briefly introduce yourself. Make sure to describe what is unique about your background and your lived experiences. Share what makes you different and what you bring to the table. Emphasize how you contribute to a diversity of perspectives and what attracted you to your field of interest. Explain from where you derive motivation and what qualities allow you to excel in your endeavors. Make it clear that you understand the mission of the institution to which you are applying, and make sure they understand why you embody their mission and how you will enhance their training program and institution overall. You can accomplish this by highlighting your unique perspectives and contributions and also what you hope to gain from joining the program and institution. It is important to customize this piece for each institution wherever possible because it demonstrates a specific interest and rationale for why they need you in their program.

RESEARCH STATEMENT

The research statement provides an opportunity for you to share what specifically interests you about your chosen field, what you plan to pursue and why, and what makes this significant or impactful. Start the essay with a "hook" about something that piqued your interest and paint a vivid picture to explain why. Summarize your educational background and describe your research experience. Only mention experiences that you would be willing to discuss in an interview. Detail what you are interested in moving forward, what you want to accomplish in your field, and why this is important on a larger scale. Include your goals and a rough idea of your career plans, even if they may change. If possible, to customize for individual institutions, state how your goals align with the mission of the institution. Also include what you will contribute to the institution and what you will gain from the training program that will allow you to advance toward your goals.

CURRICULUM VITAE

The curriculum vitae (CV) is an opportunity for you to outline all of your experiences and accomplishments. While a resumé is technically a shorter summary, this term is often used interchangeably with CV. In science and related fields, the relatively comprehensive CV is typically preferred to the abbreviated resumé.

The CV should be visually clear, verbally concise, and comprehensive without including unnecessary ancillary details. Start with name and contact information at the top. Separate sections include education with degrees earned and in progress, research experience with supervisors and funding sources, teaching experience, publications, presentations, technical skills, languages spoken/written and level of proficiency, and sometimes hobbies. Each item in these sections should have a title, a brief description only if not self-evident from the title, timeframe, and depending on the significance and your level of training may include what you did during the experience and what you learned from it. It is expected that when beginning your journey, you will not have items for all of the aforementioned categories, and you will add to them over time.

ADDITIONAL APPLICATION MATERIALS

Certain applications have a section for extracurricular activities or other meaningful experiences. The purpose of this section is generally to evaluate how well-rounded of a person each applicant is and to provide an opportunity for applicants to share important aspects of their lives. Activities listed can be seemingly unrelated to science or your chosen field of interest, but often distilled into what you did and what you learned from the experience that can then be applied to other contexts. Again, only list experiences that you are comfortable talking about in an interview or another setting.

INTERVIEWING

Do your research before interviewing for a position. Make sure you are familiar with the program's website and every page on it. Review the institution's general web pages as well. Regardless of how high on your list a specific program is, convince yourself that it is your top choice and the only place you can imagine yourself being, and go to the interview with this mindset. You can change your mind the next day, the next week, or even an hour after the interview, but you will perform better if you are entirely focused on the institution in front of you when you do the interview.

When you conduct the interview, make sure you smile, pause before answering questions, and make eye contact. Be ready to talk about anything in your application file. Prepare some examples of specific experiences you had that can be used to answer multiple interview questions. Common questions relate to a situation where there was a disagreement and how you dealt with it, a situation where you felt gratitude, an example of a time when you worked well in a team, an instance when you had to be resilient, how you overcame an obstacle and what you learned from that endeavor, etc. You can often prepare just a few examples that you can use to answer these types of questions depending on how you frame the story and the takeaway. Always choose examples of success stories; avoid red flags, such as inability to compromise, aggressive behavior, passive behavior, etc. When answering a question where you need to give an example, always make sure you understand the prompt completely, and then present your answer in the situation-action-result (SAR) format. First, briefly describe the situation (what happened). Next, detail the action (how you handled the situation). Finally, present the results of your actions and what you learned.

It can also be helpful if you prepare several updates for the interviewer if some time has passed since your application was submitted. These can range from minor to major, such as a project milestone, taking on a new leadership role, a publication, a presentation, etc. Do not say anything in an interview that you would not want anyone to hear, as you cannot predict how the information will be shared or who might overhear. Never talk negatively about another individual, institution, or field – it will reflect more poorly on you than them, no matter what negative traits may characterize the subject of such comments.

If asked whether or not the program is your top choice, we recommend indicating that it is your top choice if it is in fact your top choice. If it is not your top choice or if you are undecided, then we recommend conveying your excitement about the program and explaining that you have not finished your search or interviews yet but from what you have learned so far this is where you want to be for "XYZ" reasons and you are excited to talk to more people and learn more about the program.

After the interview, sending a thank you note to the interviewer, admissions committee, and/or director is an opportunity to express gratitude, emphasize your interest in the program, and let them know how an administrative staff member or current trainee made an impact on you. This last step allows you to elevate other people with whom you are not in competition, often setting you apart from other applicants in a very good light.

TUITION

Financing a training program can often be a limiting factor, and it is important to develop strategies for overcoming this in order to allow yourself to focus on education and enjoy your training. Obtaining a scholarship or grant is ideal, but there are not sufficient resources for each trainee to be so lucky. There are many online resources and databases of scholarships and grants by field. You can also check with foundations and academic societies that have vested interests in specific fields of inquiry. Institutional financial aid offices also play a central role in helping trainees finance their education. They can often connect individuals with low or no interest loans as well as loan forgiveness programs in some cases. If you find it necessary to work a job or multiple jobs in order to support your studies, then consider looking for student jobs where you can earn wages while learning useful skills or performing tasks that you're interested in or would otherwise be doing. For instance, many institutions hire students as tutors, TA's, laboratory assistants, etc. Another approach is to find a job where you have a lot of free or flexible time to study or do other work, such as a library or computer lab attendant.

UNDERGRADUATE DEGREE

The undergraduate experience is formative and often one of the first major opportunities to individualize in a person's life. A separate chapter is devoted to excelling in your undergraduate degree program. Here, we provide a brief review of key factors to consider when choosing an undergraduate degree program. Since the undergraduate training period is a time of exploration, it is important to make sure there are academic opportunities and resources in multiple areas of interest, even if you have an idea of what you might like to focus on. Majors and minors often evolve, and the astute undergraduate will benefit tremendously from exploring diverse topics during their training. Thus, finding a well-rounded program and institution can be extremely beneficial.

Undergraduate education is also characterized by social development. Carefully consider campus culture, living conditions, community size, and community members. Large, traditional, often public institutions usually have much larger communities with a wide array of stakeholders ranging from undergraduates to graduate students and postdocs. Smaller liberal arts colleges frequently offer a more personalized approach to education with smaller class sizes and student bodies, often entirely undergraduate. Also weigh the benefits and drawbacks of attending an often-large institution well-known for the sciences, with many resources and many parties vying for those resources, versus a smaller college with fewer resources and similarly fewer science-oriented students who often enjoy more of those resources for themselves with greater flexibility. Do you want to be a small fish in a big pond, or a big fish in a small pond?

MASTER'S DEGREE

A master's degree is not always necessary if planning to pursue a doctorate, but it can provide useful technical skills, entrée into a new domain, and may be more than sufficient for certain career goals. Some master's degree programs include an option to combine with an undergraduate degree in one extra year (4 + 1 programs) taking advantage of overlapping coursework, while other master's degree programs are standalone one-, two-, or three-year programs. Since master's degrees usually cost tuition and do not include stipends, consider the tangible skills and credibility that you will gain from the program when determining the value for your investment.

PHD DEGREE

The PhD degree, or doctorate, is an opportunity to become a true expert in a niche, dedicating over 10,000 hours toward advancing knowledge in the field of interest. Thus, entering a PhD program is a real commitment not to be taken lightly. A master's degree may or may not be required in order to enter a PhD program, depending on the field and institution. Sometimes, a master's degree is included in a PhD program as an early milestone. In rare cases, a PhD candidate can "master out" of the PhD program if they fail to make satisfactory progress toward the PhD or they choose to leave for other reasons, but this is not a guaranteed option and is generally frowned upon. Carefully consider the availability of specific mentors and principal investigators (PIs), which labs you may be able to join, and whether or not you can do lab rotations before choosing a lab for the reminder of your PhD.

Students working toward a PhD in a science, technology, engineering, or mathematics (STEM) field are often paid with a stipend or a salary based on teaching or a certain number of lab work hours. The total compensation is usually modest but should be sufficient to live on. Sometimes, larger fellowships are offered to certain candidates as a means to recruit them to the program. Also important to consider is how much teaching and/or lab work not dedicated to your dissertation project is required to receive your stipend, as these will limit the amount of time you can dedicate toward advancing your own research. Another factor in time distribution to consider is the amount and content of required and elective coursework. To gauge how productive prior trainees have been in the program, look at their publication and their subsequent positions. Also consider the track record for how long it has taken recent trainees to earn their PhD, as some programs have issues graduating candidates in a reasonable amount of time. In 2020, the national average years required to earn a PhD was 5.8 in the life sciences, 5.5 in the physical sciences and earth sciences, 5.6 in mathematics and computer sciences, and 5.3 in engineering (Kang, 2021). Since graduate students are generally the most productive at the end of their training programs and their salary is still relatively modest at that time, there is an inherent conflict of interest between the student's desire to advance to the next stage of their career and the supervisor's desire to retain the student and enjoy the fruits of their productivity for the lab group. The best PIs and institutions will support the career advancement of their trainees.

POST-DOCTORATE TRAINING

Post-doctorate, or postdoc, training may or may not be required depending on your field and your career goal. A postdoc can be completed in an academic institution, an industry setting, a government institute, or in some cases another setting. Unlike undergraduate, master's, and PhD degree programs, the postdoc is usually not a formal degree program and is arranged on a case-by-case basis with an individual PI. Thus, specific PI and lab funding will have a significant impact on your choice. Depending on the field and whether or not you have a subsequent goal position/ institution, there are benefits for doing a postdoc at the same institution or elsewhere. At certain institutions, internal postdoc candidates are viewed more favorably for faculty positions. Other times, an institution may prefer to recruit external talent to their faculty, in which case it can be advantageous to complete postdoc training at a different institution.

The postdoc training period is a good opportunity to add a new method or skillset to your investigative toolbox, or to apply a new method to a different research question. Depending on the field, a postdoc can take as little as one year, but most frequently takes four to five years (Rockey, 2012). Similar issues obtaining support for career advancement from the PI can occur near the end of the postdoc due to increased productivity, so this should be explored in advance. Since the postdoc can be the last step before obtaining your first non-trainee job, it is important that you be well poised to advance your career through the postdoc by considering individual mentors and their track record with prior trainees. Later chapters address more comprehensively approaches and best practices for choosing research mentors.

FACULTY DEVELOPMENT AND CONTINUED EDUCATION

Even after obtaining a faculty position or other non-trainee job, it is important as a scientist to continue learning. Staying up to date on cutting edge research, learning adjacent or new fields as research findings direct subsequent lines of inquiry, and faculty or career development projects are all common examples. Much of this can be accomplished by regular literature review, self-study, and attending lectures and conferences. However, there may be a place for continued engagement with coursework, as well as specific faculty and professional oriented career development programs. Many larger institutions and companies engage cohorts in project-based programs that coach individuals in larger research projects, grant writing, creation of shared resources, or other endeavors. A later chapter is dedicated to the transition to faculty and best practices of successful PIs.

REFERENCES

Kang, Kelly. (2021, November 30). *Survey of Earned Doctorates*. NCSES. https://ncses.nsf. gov/pubs/nsf22300/data-tables
Rockey, Sally. (2012, June 29). *Postdoctoral Researchers – Facts, Trends, and Gaps*. National Institutes of Health Office of Extramural Research. https://nexus.od.nih.gov/all/2012/06/29/postdoctoral-researchers-facts-trends-and-gaps/

8 Excelling in Your Undergraduate Degree

Samuel J. S. Rubin and Nir Qvit

The undergraduate degree has become increasingly overshadowed by hyped college and university experiences, skewed heavily by social endeavors. While a component of social exploration is important for one's individualization, those academically inclined will benefit from balancing this with conscious design of their undergraduate experience. In reclaiming the undergraduate degree, the savvy student will enjoy both the psychosocial benefits of college or university, in addition to the academic opportunities. Individuals who balance these domains will gain the most from their undergraduate experience and will be in an advantageous position when launching into their next role.

DESIGNING YOUR EXPERIENCE

Thoughtful design of the undergraduate experience necessitates the student take an active role in the process. The well-balanced undergraduate will engage with a variety of extracurricular activities, coursework, teaching opportunities, and research throughout their tenure in a harmonious manner. It is important to seek and consider different opportunities, ask advisors, faculty, and senior students for guidance, read online materials, and comprise a mental log of the general options. Without investigating, one can easily fail to benefit from the unique opportunities and resource-rich environment in which the undergraduate may explore. We recommend maintaining an open mind, trying new things, considering taking a course in a new field or going to a lecture or group meeting outside your comfort zone. Enhancing your exposure broadens your horizons, improves your creativity, and expands your networks. While this process of exploration and learning is tremendously beneficial, it is also important to avoid over committing yourself with too many engagements such that you cannot meaningfully immerse yourself in any one of them. No individual can be in all the clubs, take all the courses in a department at once, or be on all the sports teams. It is far better at the beginning to learn a little bit about a lot of things before diving in deeply. Then, make a thoughtful choice about which opportunities and activities to pursue in more depth. The best approach is to under promise and over deliver.

Most undergraduate institutions offer a vast array of resources if you look in the right places. These assets range from advising and career services, to creative spaces with tools and raw materials, free software and hardware, free transportation,

DOI: 10.1201/9781003301400-8

unusual study spaces, alumni networks, etc. Take advantage of these resources while you have access to them. Whether you are paying tuition or not, you deserve these resources – they are set up for students to utilize.

Some of the most important and often underutilized resources come in the form of human intelligence. At an academic institution, there are numerous individuals that may be able to offer general advice, experiential wisdom, academic reputation, and professional networks. Seek help and guidance whenever useful and as often as possible. You don't know what you don't know until you ask. Practice curiosity and investigative techniques. You will save time and end up ahead of your peers without reinventing the wheel.

EXTRACURRICULAR ACTIVITIES

A significant portion of undergraduate education takes place beyond the scope of structured academic pursuits. This takes the form of extracurricular activities, which can range from unstructured academic time, to creative, volunteer, social, and/or physical activities. These extracurricular spaces are integral to the process of exploration, individualization, and ultimately holistic education. The most important guidance we can offer is to have fun with this unstructured time. Take up new interests. Learn new things. Explore. Whether joining a chemistry club, a robotics team, a ceramics studio, tutoring, a salsa dancing class, volunteering at a food bank, playing on a sports team, learning to rock climb, or going hiking with friends, take advantage of your unstructured time to build balance in your life and offset the time spent on structured academic study. Keep in mind that you also don't need to do everything all at once just because it's available. Spreading yourself too thinly across many activities can have the opposite effect and instead lead to exhaustion and dropping the ball. Experiment by adding one or a few activities at a time, see what you have time for, switch to something else, and continue the activities that you enjoy and find time to engage with. You are under no obligation to continue going to every meeting of the urban horticulture club just because you went to the first one (I did not).

A NOTE ON STUDY ABROAD

Study abroad is another opportunity that you may come across at your under-graduate institution. Generally, this involves an established exchange program with a university in another country where you can go for a quarter or semester or year to take courses and live in a new place. This type of program often represents an appealing chance to travel. Spending time in another country can certainly help master a new language if you are able to develop a foundation in that language before traveling and then fully immerse yourself in the language while you are in that country. Study abroad also allows you to build a connection and network in a particular location or institution of unique resource. Beyond these scenarios, study abroad programs usually do not turn out to be the most academically rigorous despite their hype or what you hope will be possible. With some exceptions, major academic progress or accomplishments during this brief period away from one's home institution are not common. Since the durations of study abroad programs are relatively short and much of the time is spent on relocation, learning a new system, and the excitement of being in a new place, these opportunities represent much more of a social exploration than opportunity for deep academic engagement. That being said, fun social exploration is important, and electing to do a study abroad program is perfectly acceptable if you have made sufficient progress academically such that this type of academic disruption is unlikely to set you back. Overall, we generally recommend against study abroad programs with the exception of intensive foreign language study. Outside of this scenario, it is probably most efficient to focus on academic study at your home institution during the term and then focus fully on fun and travel with little or no academic expectations during your vacation or break periods.

COURSEWORK

Coursework is obviously one of the center pieces to undergraduate educations, and there are several strategies worth considering in order to make the most of this time. Institutions have varying degrees of general class requirements as well as major-specific class requirements. It is important to be aware of these requirements up front, and it can be helpful to complete them earlier rather than later so that you can establish a strong foundation and have more time to explore electives and advanced courses once you get your bearings and have a better idea of what you want to pursue in the greatest depth. If you took Advanced Placement (AP) and/or International Baccalaureate (IB) courses in high school, you can often receive credit toward general undergraduate requirements, which will free up your time to take other courses of interest and may also give you certain priorities during registration based on total credits accrued (I got to live in a senior dorm during sophomore year due to AP and IB credits). Don't be afraid to petition the registrar or administration to count a different class toward a general requirement if there is a compelling rationale to do so. I once fulfilled a first year writing seminar by taking a course on composition of laboratory reports.

Also make sure you have some time to explore potential interests early on so that you can affirm your desired trajectory and ensure that you are completing the appropriate requirements. If you switch majors, it would be a shame to have just completed the introductory requirements for a different path, only to start over again in a new field (this happened to me even later on). Of course, this is sometimes unavoidable and a natural part of the exploration process – but generally best avoided when possible.

When considering which specific courses to take and with whom to take them, ask senior students and consult online resources that contain prior syllabi or professor ratings. See if you can shop around multiple courses at the beginning of the term and then retain those which are most interesting to you and appear to have the best learning environment for your needs - based on the professor, teaching style, course structure, class size, grading rubric, etc. In general, specific grades don't matter as much in undergraduate education compared to high school. The main distinction is between an A and B student versus a C and D student, and undergraduate work products such as written compositions and presentations speak louder than grades. Since the number of courses you can encompass during an undergraduate career is far more limited than all those available in the course catalogue, we recommend "spending" a course on something that you will best learn in a classroom environment, with a professor, classmates, and a prescribed curriculum. If you can learn a skill via reading or online video, then do so and utilize a valuable course in your schedule for taking your knowledge to the next level in a manner not possible via self-study. If a computational biology course requires introduction to Python as a prerequisite but you already learned Python yourself, don't hesitate to explain this to the professor and ask to demonstrate your prerequisite knowledge in order to jump directly into the advanced applied course you want to take.

TEACHING AND TUTORING

Teaching and tutoring are great opportunities to reinforce your learned knowledge and also demonstrate a commitment to helping others in general or in a specific field of interest. Many institutions maintain peer tutoring programs, community tutoring programs, and advanced teaching assistant (TA) opportunities. Some positions are paid, while others are volunteer based. Regardless, we find these opportunities mutually beneficial to the teacher and the learner. Both parties generally derive significant gratitude from the experience. We recommend seeking out opportunities to tutor younger students in the community to start. When you have taken some courses and built a strong knowledge base, peer tutoring can be an excellent method to solidify and maintain your knowledge. Once you identify a particular are of focus or interest, serving as a TA for a course (usually a course you previously completed) is a coveted and honorable opportunity in leadership and teaching. We find that if we cannot teach a concept, then it is difficult to establish our own understanding of that concept.

RESEARCH

Engaging in research early during the undergraduate tenure is an extremely advantageous choice when considering a career in any aspect of the sciences. Doing research demonstrates initiative and serious interest, and the transferrable skills learned are invaluable to any future endeavors. Getting your feet wet early allows you to start learning the scientific method in practice and allows you to hone your scientific critical thinking. Moreover, it is helpful to determine as early as possible if the research career is a good fit for you, and in many instances, undergrad will be the first exposure to real scientific research. The availability of research opportunities and research mentors depends on the setting. Undergraduate only colleges present ideal environments for students to work directly with professors and have the greatest responsibility for their research projects. Larger universities with graduate students and/or postdocs also present their advantages for the undergrad looking to work with these advanced trainees. However, in the latter environment it is rare for the undergrad to have as much responsibility and ownership over the research project since there are more advanced trainees involved.

Sometimes, there are opportunities to volunteer in a lab doing research during the academic term, or to get paid for this on an hourly basis. It can also be very informative, formative, and productive to spend a portion of the summer break doing research on campus or at another institution. There are a number of ways to obtain funding for this type of position. Many institutions have internal grants or fellowships to support students doing research during the summer or during certain terms. The National Science Foundation (NSF) and other private foundations or companies also offer undergraduate research fellowship programs, most often during the summer break time. These fellowship programs are excellent, structured opportunities to gain exposure to research and broaden your network in the scientific community. These positions also usually come with some minimal associated curriculum, which can be helpful for learning related research skills or preparing for graduate school or job opportunities.

The undergraduate thesis is one of the first opportunities for a budding scientist to contribute a new piece of knowledge to the field. Some theses are based upon literature review, while others will include novel experiments and findings. We recommend attempting to pursue the latter whenever possible, as learning this process early will give you a head start in your subsequent training. Take this opportunity to experiment with the scientific method, discuss with expert advisors, make revisions, and disseminate findings. The new piece of knowledge contributed can be small, but the process is nevertheless a significant and meaningful learning experience.

PLANNING NEXT STEPS, GRADUATION, AND TRANSITION

Undergraduate education will come to a close before you know it, and it is important to plan early for your next steps. A good time to start this planning for this is the summer before your final year (or even a year before that if you are interested in medical school). Scope out opportunities of interest – whether post-baccalaureate training programs, graduate degree programs, and/or job opportunities. Request letters of recommendation early, understand what applications require, and start drafting your materials in advance so that you have time to ask mentors and advisors for feedback on how to improve the competitiveness of your candidacy. Submit your applications early, and celebrate your accomplishments. These milestones don't come every day, and it is important to give yourself credit for your hard work. Take advantage of transition time between undergrad and your next position. Travel or do something fun that would not otherwise be possible, and embark upon your next step with a fresh sense of inspiration.

9 Postbaccalaureate Programs

Samuel J. S. Rubin and Nir Qvit

A postbaccalaureate program (postbacc) is a continuing education program designed for individuals who have completed an undergraduate degree and are interested in applying to a graduate degree program, and either need to gain more experience in order to inform their decision or need to fulfill prerequisites in order to become a more competitive or eligible applicant. Postbacc programs usually last one to two years and serve a transitional purpose. A certificate may be given, but a postbacc program is not an advanced degree. Thus, postbacc programs should be thought of as an optional steppingstone, not a specific requirement and not a standalone form of training.

There are several different types of postbacc programs that serve distinct purposes. A postbacc program is usually geared toward preparing individuals to pursue advanced graduate degrees and subsequent careers in either scientific research or medicine. Some of the research-oriented postbacc programs are broad and inclusive of experience in biosciences, biotechnology, or other areas; others are focused specifically on chemistry, biology, physics, or other niche fields. These research-oriented postbacc programs are centered around a mentored experience in a research laboratory where the trainee learns basic lab skills and gains research experience hands on. Frequently there are additional optional or required foundational lectures and career development seminars. Resources may also include preparation for graduate program entrance examinations, such as the Graduate Record Examinations (GRE), as well as graduate degree program application coaching. In developing relationships through these experiences, there are also often opportunities to obtain letters of recommendation and additional support in transitioning to the next step in one's career. Pre-medical postbacc programs are designed to prepare individuals who plan to pursue medical school or other health profession degree programs and often contain more coursework and no research or less research than the research-oriented postbacc programs geared toward subsequent PhD programs. Most postbacc programs are offered by academic institutions, although some can be found in industry (*e.g.*, pharmaceutical companies) or government (*e.g.*, National Science Foundation (NSF) or National Institutes of Health (NIH)).

The main benefits of a postbacc program are the resources provided to trainees and the opportunity to utilize those for strengthening one's application to graduate degree programs. Developing a longitudinal mentorship with a principal

DOI: 10.1201/9781003301400-9

investigator in a research-oriented postbacc programs allows for personalized career guidance and support, as well as a strong letter of recommendation. The lab experience also allows trainees to learn a variety of basic skills, which become marketable expertise for graduate school applications. In addition, trainees gain an insider understanding into the functioning of a successful research laboratory if chosen wisely. Coursework offers the opportunity to expand foundational knowledge and assimilate new domains, as do seminars and lectures. Test preparation courses or materials can provide valuable accountability and practice for standardized examinations required by graduate degree programs. Career development workshops are often included to help trainees optimize their resumes, application materials, and even practice for interviews. Also consider whether the postbacc program includes a stipend or requires tuition. This can vary across fields and institutions, but generally research-oriented postbacc programs are compensated, while pre-medical programs charge fees. Most of all, the postbacc program allows an individual to buy extra time to make an informed decision about their next steps while enhancing the strength of their candidacy for whatever that next step may be.

Pursuing a postbacc is most appropriate when it fulfills one or both of two purposes. The first purpose is to gather additional experience and information in order to better inform your decision-making capacity regarding what specific field or type of degree to pursue next. In this scenario, pursuing a postbacc program will provide additional time, resources, and exposure in order to gain clarity about desired career trajectory. For instance, it can be helpful to determine before applying to and entering a graduate degree program whether you prefer "wet lab" bench research or "dry lab" computational work, what field or topics are most exciting, etc. The second appropriate use of a postbacc program is to fulfill core prerequisites in a new field or in greater depth than possible during undergraduate training. For instance, this may include prerequisite coursework and/or simply additional laboratory research experience and applied expertise. The goal should be to become a stronger applicant when applying to your desired subsequent training program. Since the postbacc program model serves a transitional purpose, it is not appropriate to use as a terminal training program or to gain a degree commodity. If your goal is to advance your knowledge in a specific identified area before pursuing another graduate degree or as a destination experience, then a master's degree program may be more appropriate.

10 The Master's Degree

Nir Qvit and Samuel J. S. Rubin

INTRODUCTION

Master of Science (MSc) degrees can be traced back to medieval institutions, with universities emerging all over Europe. Although these universities primarily concentrated on theological and philosophical studies, some scientific inquiry and discovery flourished during the Renaissance and Enlightenment periods. This laid the groundwork for MSc degrees as specialized qualifications in scientific disciplines, primarily in the natural sciences and mathematics. University graduate programs have been established since the 19th century in order to provide deeper training in a variety of scientific areas (*e.g.*, physics, chemistry, and biology, etc.). With the expansion of scientific knowledge, MSc programs expanded to include a wide range of disciplines (*e.g.*, engineering, computer science, biomedical informatics, biotechnology, environmental sciences, social sciences, etc.). Recently, many institutions have offered either highly specialized tracks (*i.e.*, professional-oriented courses emphasizing practical skills) or more flexible programs (*e.g.*, interdisciplinary courses), allowing students to tailor their studies to their specific interests. A number of programs incorporate practical elements as well (*e.g.*, internships and industry collaborations).

The MSc degree program has gained increasing recognition and importance in academia, business, and industry. There is a great deal of value placed on graduates of these programs by employers throughout the world since these individuals possess advanced knowledge in a wide range of fields depending on the focus of their specific program and are capable of becoming scientists, engineers, data analysts, consultants, or educators in the academic, industrial, and research fields. Since these individuals have developed expertise in a particular scientific field, they can advance knowledge, drive innovation, and solve complex problems. Thus, many industries actively recruit graduates with master's degrees.

ADVANTAGES AND DISADVANTAGES OF AN MSC DEGREE

There are many benefits of earning an MSc degree for individuals who want to advance their education and careers. The MSc degree provides an opportunity to acquire advanced knowledge and specialization in a specific field, gaining a comprehensive understanding of the subject matter. There is also considerable evidence that advanced degrees enhance career prospects since employers often prefer candidates with advanced degrees and may even require these qualifications in order to fill certain positions. Thus, master's degrees can provide the opportunity to

DOI: 10.1201/9781003301400-10

obtain higher-level jobs, increased responsibilities, and higher salaries. The professional networking aspect of this mobility is also an important aspect to keep in mind. By making contact with higher level academic and industry professionals, one can receive mentorship, engage in collaborations, and gain further professional opportunity and mobility. Last but not least, in the course of master's programs, trainees may be challenged to acquire critical thinking skills, conduct research, and solve problems. As a result, students grow and develop as individuals, adapting to changing environments, and increasing their knowledge and abilities to analyze complex issues.

There are several disadvantages of MSc degree programs that are worth noting as well. The pursuit of an MSc degree involves a significant investment of time (usually about two years, although occasionally one or three years) and money, since typically students cannot work (at least not full time) during this period and there are many expenses associated with the degree (*e.g.*, supplies and tuition fees – unlike PhD programs with stipends or TAships). Some individuals take out loans to fund their enrollment in these programs. If finances are a limiting factor, it is therefore vital to carefully assess the likely return on investment in terms of time and money required to earn the master's degree (as well as missed opportunity during this time) versus potential future higher salaries or other gains after completing the degree. Occasionally, advanced degrees may also lead to over-qualification, since some employers or job roles may not require such a high level of education and may prefer to hire individuals without these degrees at a lower salary. Not all fields require a master's degree either. Therefore, it is extremely important to make sure that the degree chosen is aligned with the career goals and professional development of the individual.

MASTER OF SCIENCE (MSC) VERSUS DOCTOR OF PHILOSOPHY (PHD)

An MSc degree is a relatively short program designed to provide students with advanced knowledge and skills in a specific area of study. In general, a master's is considered a more coursework- than research-based degree. Alternatively, a PhD is the highest academic degree one can attain in most fields and involves some coursework plus teaching and a research-intensive program conducted over four or more years. In order to achieve a doctorate, one must conduct original research and make significant contributions to the field. In some fields or countries, a master's degree is required before enrolling in a PhD program, or the degrees can be obtained sequentially in a combined program. Both degrees are considered to be advanced training, but the PhD program is generally regarded as a more rigorous and specialized qualification that represents advanced academic achievement. As a result, PhD holders have higher and more diverse employment opportunities in academia and in industry, where in-depth expertise and the ability to conduct independent research are highly valued. On the other hand, an MSc degree provides the opportunity to gain advanced expertise in a new or ongoing field of study without as much time required as a PhD degree. MSc degree holders may pursue careers in industry, academia, or government at entry-level positions. In pivoting to a new field, a master's degree may be a useful opportunity to test the waters before taking on a longer commitment in that field and can also provide additional qualifications should you wish to pursue next steps in the field. In addition, some individuals with experience or an advanced degree in one field may choose to pursue a master's degree in another field in order to gain experience and qualifications in a method or body of work with the goal of applying that knowledge to their primary field of interest. Thus, a master's degree can serve a variety of purposes in place of or in addition to a PhD or other graduate degree.

MSC DEGREE PROGRAMS IN THE UNITED STATES OR EUROPE

There are many reasons why an individual may choose to pursue an MSc degree in the United States (US) or Europe. Both regions have universities with long-standing traditions of rigor and innovation in academics. The decision should be made carefully by considering the individual's goals, preferences, and circumstances. Generally speaking, each region has its own advantages and disadvantages. For this reason, it is crucial that an individual studies each program's benefits and shortcomings to make an informed decision before enrolling at a specific university and pursuing a particular career path within a field of study.

The United States and Europe each have many prestigious universities with internationally recognized programs known for their research outputs and academic excellence. Some of these institutions include Stanford, Harvard, and MIT in the United States, as well as Oxford, Cambridge, and ETH Zurich in Europe, just to mention a few. These universities are globally renowned for the quality of their research and programs. Nevertheless, it is important to verify the rankings and academic or professional reputations of specific programs in your field of interest to make an informed decision about what benefits you may derive from the degree, which is not without financial and opportunity costs. In general, an MSc in the United States is considered to be a more flexible program (*e.g.*, the choice of courses and the ability to customize the curriculum), which allows students to plan their curriculum according to their interests both academically and professionally. In contrast, the European education system has a less flexible structure, with set curricula and limited opportunities for electives.

In today's world, advanced degree programs are a great way to establish a professional network and form scientific collaborations. In both the United States and Europe, students are encouraged to create networks of alumni with strong connections to industry in order to succeed in the job market. Because of this, network building locally can be very important for job seekers, unless they are applying to global companies. In light of this, it can be particularly helpful to evaluate career prospects and alumni networks prior to applying to any program or institution. By assessing which networks align with your career goals, you will be able to make an informed decision.

MSc program tuition fees are generally higher in the United States than in Europe. In the United States, there are many funding opportunities available for students (*e.g.*, scholarships and loans), and some institutions even offer financial aid packages to students in order to help support them while they attend training. In spite of the fact that European universities have lower tuition fees, supplementary funding options are limited.

Study in the United States necessitates a strong proficiency in the English language. This allows students to fully engage with coursework and academic networks. US programs also expose individuals to a diverse student body, which will provide the opportunity to get to know a vibrant and multicultural society. Many European countries offer English language programs, some of which may also require students to have proficiency in the local language. Therefore, options can be limited in Europe depending on individual language proficiencies. On the other hand, there are also significant benefits to learning new languages if able, gaining a more global perspective and network, and immersing oneself in different cultures.

Ultimately, the decision regarding whether or not to pursue an MSc degree, and if so whether to consider programs in the United States or Europe or elsewhere, depends on individual circumstances, preferences, and goals. The United States and Europe both have exceptional programs available. Consult with professors, professionals in your field, and current students as with any other academic or career decision.

CONCLUSION

The MSc has shaped scientific training and prepared many individuals for successful careers in their chosen fields. As we continue to push the boundaries of knowledge, the MSc degree remains a symbol of intellectual achievement, even as the field of science continues to advance and more trainees go on to earn doctorate level degrees. While the MSc generally requires more financial investment in tuition than a paid PhD degree, it requires less time investment than a doctorate level program and provides many advantages to individuals seeking advanced training in a specific field. MSc degrees may be pursued prior to, in conjunction with, or in place of PhD degrees, depending on individual goals. Master's degrees offer high-level knowledge, career advancement, potential job opportunities, professional networking, and personal growth to those who seek this in a one- to three-year advanced degree program.

REFERENCES

Burkhardt-Holm, Patricia, and Camelia Chebbi. "Master's Degree in Sustainable Development in Switzerland, the First Master Course Comprising Three Faculties." *Environmental Science and Pollution Research* 15, no. 2 (March 1, 2008): 136–142.

Cohen, Richard, Lucas Murnaghan, John Collins, and Dan Pratt. "An Update on Master's Degrees in Medical Education." *Medical Teacher* 27, no. 8 (December 1, 2005): 686–692.

Ullrich, Charlotte, Cornelia Mahler, Johanna Forstner, Joachim Szecsenyi, and Michel Wensing. "Teaching Implementation Science in a New Master of Science Program in Germany: A Survey of Stakeholder Expectations." *Implementation Science* 12, no. 1 (April 27, 2017): 55.

11 Excelling in Your PhD

Samuel J. S. Rubin and Nir Qvit

Completing a PhD is a massive undertaking; the process is also a special chance to become an expert in a specific field and to discover and contribute new knowledge to the community. In order to be successful in the PhD, one must understand what the commitment entails and be prepared to approach the associated challenges thoughtfully and strategically. In this chapter, we address opportunities and common pitfalls likely to be encountered with the PhD.

CHOOSING THE RIGHT PROGRAM

Finding the right program for you is the first key step toward ensuring success during your PhD. Chapter 7 in this book focuses on identifying and applying to the most promising training programs for your interests. In brief, a broad and systematic approach to identifying programs will help ensure that you apply to those that represent the best potential fits. Following best practices during the application and interview process will help ensure that you receive more offers of admission. Herein, we aim to provide specific advice on choosing the best PhD program for you.

There are several unique factors to consider when choosing a PhD program, and they should be assessed according to your specific priorities and goals. Some of the attributes most relevant when comparing PhD programs include selection of laboratories accepting new PhD students; availability and accessibility of principal investigators (PIs, usually professors who serve as primary advisors and lab directors for trainees) who could serve as dissertation advisors; availability of laboratory funding; availability of additional institutional grants to support early stage research projects; amount and content of required coursework; availability of elective coursework; required and optional teaching opportunities; whether a stipend is guaranteed by the program/department or contingent upon PI funding, teaching, or some other requirement and for how long it is guaranteed; responsiveness of the PhD program to student feedback; opportunities to work with undergraduates, master's students, postdocs, or other trainees; institutional resources and core facilities available to PhD students at the institution; average length required for recent PhD students to complete the program; types of post-graduate positions obtained by recent graduates (*i.e.*, academia, industry, government, etc.); academic reputation of the program, department, or institution. Certainly, many additional criteria should be

DOI: 10.1201/9781003301400-11

evaluated when choosing the best PhD program for your goals, but these are some of the key factors especially relevant to the PhD degree.

Particularly prudent to consider are (1) the research mentors you would be working with and the kind of research you could conduct; (2) guarantee of funding for your stipend and your research; and (3) the post-graduate destinations of recent program graduates. Since the most significant portion of the PhD is based in research and much of your learning is derived from your PI and the environment they create in their lab, it is critical that this be a central consideration to your choice of PhD program. Make sure that there are multiple laboratories in which you could envision working before choosing a program, as sometimes a PI or lab will move to another institution or lose funding for new trainees. As your ability to conduct research is dependent on protection of your time to do so as well as funding of resources required for experiments, it is also imperative that you verify the source of PhD student stipends and research funds and that these are likely to sustain your PhD tenure. Lastly, it is also important to ensure that the PhD program will embrace and support your career goals and that there is a track record of prior trainees obtaining subsequent positions consistent with your goals. Even if you don't know whether you want to pursue an academic, industry, or other career, it is important that the program support that ambiguity. Many programs expect all their graduates to pursue academic careers, which is not an issue if that is your goal but can obviously be problematic if you have other or unclear aspirations.

THE BEGINNING

Setting the stage for the PhD should involve cultivating a mindset and expectations consistent with the reality. The PhD is a marathon, not a sprint. Do not rush to the detriment of quality or endurance. One must build stamina to prepare for the journey. You will learn a way of thinking. You will learn to teach yourself anything. You will become an expert in a niche field, make discoveries, and share those with the field. Perform extensive due diligence when joining a lab or starting a project to align your expectations with reality and make an informed decision whether a given path is right for you. Dedicate effort to creating a strong foundation and community at the beginning of the PhD program to support you through this significant undertaking. Build relationships with other students, with affinity groups, and with local communities. You will rely on these outlets for support in times of stress and success.

THE STUDENT-ADVISOR DYAD

The student-advisor relationship is a sacred, central aspect of the PhD, shaping research experience and career trajectory at all levels and for years following completion of the degree. Choosing an advisor (synonymous with PI and lab for this purpose) is a career defining decision, and should be approached strategically, methodically, thoughtfully, and carefully in order to ensure a successful outcome. Care must be taken to nurture the student-advisor relationship at all stages of the PhD, which ideally involves taking an active role in directing the partnership. The student-advisor relationship is unique because although there is inherent superiority of the advisor to the student due to academic hierarchy and years of experience, there is also an interdependence and partnership that characterizes the relationship. The advisor must take on responsibility for guidance, teaching, and support of the student, and the advisor is simultaneously dependent on the student to carry out research and advance the lab's mission. Thus, the student-advisor relationship is a special partnership that requires conscious design.

Great care should be taken when choosing an advisor and lab for the PhD, given the significant impact this has on your entire trajectory and the challenges associated with switching advisors/labs (see Chapter 20 for a discussion on switching labs). Some PhD programs offer or require laboratory rotations at the beginning of the program, which we highly encourage as an excellent way to try out an advisor/lab and also to learn new skills in different domains that you may later wish to apply to other research questions. Regardless of whether or not you are able to do rotations, we suggest a systematic approach to evaluating potential advisors/ labs. Just as you searched for PhD programs, consult your network to identify promising leads, and do peruse the full faculty roster for the institution online. Look for PIs either doing research in a field of interest or applying concepts and methods of interest. For labs of interest, reach out to the PI via email with your CV to schedule an introductory meeting. This two-way interview begins the courtship process, in which both you and the PI evaluate whether your partnership would be a good fit. Ask and consider the following factors: availability of research projects; openness to new research projects proposed by trainees; abundance and timelines of current research funding (verify this on the NIH or other funding body websites); lab resources (equipment, cell lines, animal models, etc.); physical lab space; lab culture (social, remote work, collaborative vs. individual projects, etc.); expected work hours; who writes funding applications; who writes manuscripts and decides when and where to publish papers; mentorship style (frequency of meetings, hands on vs. hands off, etc.); support for attendance and presentation at conferences; how it is decided when PhD students are ready to defend their dissertations and graduate; how long recent PhD students required to complete their degrees; support for your career goals (academic, industry, etc.); to what extend the PI advocated for recent graduates in obtaining their subsequent positions; professional success of former trainees; whether the PI has tenure; and whether the PI plans to stay at the institution for the next five or more years. Although they may seem forward, it is perfectly acceptable to ask these prudent questions. Doing this diligence demonstrates genuine interest, and a sophisticated PI will appreciate this effort. If your questions are evaded, this raises red flags. Also make sure to consult with current and former trainees of the PI to understand their experiences, good and bad.

The student-advisor relationship is most fruitful when actively cultivated throughout the PhD. According to both the student's and the advisor's desired level of engagement, time should be set aside for regular meetings to discuss and troubleshoot research, consider career goals, plan for publications and conferences, and coordinate additional learning and enrichment opportunities. Ideally both parties make a continual choice to partner with one another. The relationship is usually a mix of personal and professional in nature, as it is a unique longitudinal and intensive intellectual mentorship connection not found in many other domains outside academia.

Like any positive relationship, open communication and proactive expectation setting by each party are key components to success. At times, this can require a fair amount of assertiveness on the part of the student, which can be challenging given the academic power differential. Students often need to advocate for meeting time with busy advisors, guidance with preparing and submitting manuscripts for publication, and advancing toward other milestones. Thus, it is all the more important to communicate and set these expectations at the beginning of the relationship and perhaps even put those in writing, reaffirming and adjusting as needed. (Simply writing a follow up email to a meeting outlining what was discussed serves as a sufficient reference point for both parties to avoid subsequent miscommunications.)

A permutation of the student-advisor dyad is co-mentorship, which is becoming increasingly common although still represents the minority of cases. In a co-mentorship, the student selects two advisors. This arrangement is most appropriate when designing a project that applies some specialized method to a different field, in which case one advisor serves as an expert in the method and the other advisor serves as an expert in the applied field. Generally, the co-mentorship relationship functions best when there is one primary mentor most responsible for the student on paper for bureaucratic and career advancement purposes. To be successful, a co-mentorship also requires careful goal setting, expectation alignment, and continual communication. A scenario in which the student ends up conducting two parallel projects in different laboratories should ideally be avoided because this detracts from the ability to focus deeply on one niche. Alternative structures that support multiple mentors without splitting effort into a full co-mentorship include collaboration and advisory committee membership. As a member of one research group, it is very feasible to collaborate with another lab for the purpose of engaging a new technique or field application, thus avoiding the complexity of a formal co-mentorship. Another option is to invite such PIs to join the thesis committee or other advisory committee for the purpose of receiving their mentorship and guidance. These excellent alternatives help maintain the more straightforward student-advisor dyad without the complexities of co-mentorship but with the opportunity to engage additional peripheral mentors.

An extension of the student-advisor relationship involves the assembly of the thesis committee, which includes additional advisors and mentors that help guide students through the PhD. It is important to choose advisors for the committee who will both provide helpful guidance and appropriate critique, while remaining strong supporters of your academic progress and success. Given that these committee members influence or determine passing of the qualifying exam and advancement to PhD candidacy as well as passing of the dissertation defense and graduation, it is critical to avoid any committee members that could stall or halt your advancement. It is better to seek out informal advice from such individuals than to place them on a committee with such control over your advancement.

NAVIGATING ACADEMIC POLITICS AND ACADEMIC MANNERS

Politics and manners in academia share some features with corporate culture, but there are also important differences. Academic hierarchy typically yields to level of education and seniority. Many academicians are curious, skeptical, and collaborative, while some are infamously egotistical and hungry for control over direction of their field. In general, it will benefit you to hope for the best intentions when interacting with others and guard against worse intentions. Communicate openly and collaboratively, but do not feel pressed to share more information than comfortable. Always assume that what you say may be relayed to anyone else, no matter how much you trust the individuals in your presence or what assurances they offer that the information shared will be kept confidential. Demonstrate your stature by thanking people for their patience rather than apologizing for your delay in following up. Whatever you put in writing (email, text message, etc.) will become part of the record for better (documenting key details) or for worse (evidence of negligence or wrongdoing).

Difficult conversations are sure to arise in anyone's career. These may relate to disagreements, adversarial relationships, or simply uncomfortably topics. I took a very useful course on this topic at the Stanford Graduate School of Business called "Managing Difficult Conversations" that was taught by William F. Meehan III, MBA (Lecturer in Strategic Management and Senior Partner Emeritus of McKinsey & Company) and Charles G. Prober, MD (Professor of Pediatric Infectious Diseases and former Senior Associate Dean for Medical Education). In this course, Mr. Meehan and Dr. Prober taught their guiding principles of a difficult conversation, which include: assure confidentiality; be non-judgmental; do not mislead; emphasize what you know; show empathy; do not mistake vagueness with compassion; keep language succinct and simple; offer hope and comfort; be calm and calming; be an active listener; pauses are your friend; and remember it is *how* you say it. These principles can be deconstructed into a sequence of steps to conduct successful difficult conversations, according to Mr. Meehan and Dr. Prober. Before the conversation, identify clear goals, assess the recipient's emotional state, prepare an outline, find your voice, anticipate questions, and practice. Begin the conversation by choosing the setting, stating the goal clearly, getting directly to the point, addressing information asymmetry, and following a logical approach. Conduct the conversation by asking open-ended questions, checking in frequently, using frequent pauses, outlining next steps, and reflecting immediately after. This course included an extensive component of active practice, and I strongly suggest that you seek a similarly structured learning opportunity to prepare for these inevitable situations.

SUBSTANCE OF THE PHD

As discussed, the major substance of the PhD is based upon novel research. You will live, breathe, eat, sleep, dream, and exude science. In subsequent chapters, we discuss in detail major tasks, such as designing a research project (see Chapter 13), obtaining funding for research (see Chapter 14), conducting excellent research (see Chapter 12), publishing and presenting scientific findings (see Chapter 15), and more. We recommend considering strategic approaches to these endeavors, as there are many common pitfalls and mistakes.

LITERATURE REVIEW

One of the major parallel efforts intertwined with research and often overlooked or underprioritized is getting up to speed with the current state of your field and staying up to date with cutting-edge research through the primary literature. Familiarizing yourself broadly *and* deeply with historical research articles and review papers on topic generally requires months to years of reading, providing a strong foundation and understanding of common techniques utilized in the field and approaches to problems already tried. Simultaneously and subsequently perusing all the current literature as it is published will allow you to stay up to date with cutting-edge methods, discoveries, and applications in your field, as well as the most influential individuals and laboratories working in the area. This is important for inspiring novelty in your work, avoidance of "reinventing the wheel," and identifying potential collaborators and competitors. Therefore, it cannot be emphasized enough how important it is to dedicate significant effort and time on a regular basis to becoming intimately familiar with the primary literature in your area of focus. Early in the PhD, this will consist of several hours of reading daily – both past and present studies, which will help inform your research plans. Later, your reading efforts should involve reading several current research articles per week most relevant to your field, thereby maintaining the forefront of your knowledge while allowing you to focus most of your time on your own research. Create PubMed or other alerts based on search criteria relevant to your niche so that you receive regular notification of new studies, and conduct systematic literature reviews when assimilating a new or adjacent topic.

Identify respected sources in your field of interest, but keep in mind that higher impact "rigorous" journals also have higher retraction rates due to the pressures and politics of publishing there. Read critically and develop a sense of how much to trust the methods, the data, and the interpretations of each study. Maintain skepticism. Pay attention to authors' training and affiliations, as well as sources of funding, which may reveal more than the declared conflicts of interest. Ask yourself whether the authors' conclusions can be supported by the data presented, if sufficient data has been presented to convince you of the conclusions, whether you are confident in the reliability of the study's methods, and whether there are any additional factors (*e.g.*, sources of funding or other competing interests) that cast doubt on the validity of the study. These key features contribute to the provenance of each study and are just as or more important than the results themselves. Without provenance, results lack meaning.

LABORATORY SAFETY

There are far too many instances where students or other laboratory personnel were injured, or worse, while conducting experiments. There is an infamous incident (and criminal case) relating to inadequate use of a lab coat that took place at UCLA, resulting in a tragic fatality, which sparked widespread changes in lab safety practices. Our philosophy can be boiled down to the notion that if you do not have adequate and appropriate personal protective equipment (PPE) and knowledge of how to use it properly, then you should not be doing the experiment. It is really that simple. PPE can range from hoods, ventilation systems, shields, lab coats, gloves, goggles, masks, respirators, and other supplies. It is inappropriate for anyone to pressure you or someone else to perform a task that could endanger your safety or the welfare of others and for which you do not have adequate PPE to carry out the task safely. Cost or inconvenience of adequate PPE is not an excuse to endanger personnel.

LABORATORY SUSTAINABILITY

Since most reagents are limited in supply and costly to obtain, it behooves each investigator to be mindful of the sustainable use of materials. Do not compromise the validity of experiments by attempting to reuse supplies that should be discarded since ultimately this will cost more resources in repeating the experiment when gone wrong, but do be thoughtful about avoiding unnecessary waste and energy use that depletes funding sources and negatively impacts the environment (much of laboratory waste is incinerated). Store stocks and reagents with labels in proper containers and locations to protect them from spoilage, misplacement, contamination, negligence, and malfeasance. Make aliquots of commonly used reagents, label them, and preserve them appropriately (protected from light, temperature, etc. as needed) so that individual aliquots can be retrieved for specific experiments without contaminating or spoiling the rest of the stock. This small bit of extra effort up front will ultimately save valuable time and money for other experiments.

APPROPRIATE ANALYSIS AND INTERPRETATION OF DATA

In order to analyze and interpret data appropriately, it is tremendously critical to have a deep fundamental understanding of statistical methods. Far too often is this overlooked for a superficial notion of when to use each of a few common statistical tests. Without a deeper understanding of at least the fundamental concepts of statistical inference and basic statistical hypothesis testing, it is far too easy to inappropriately design experiments, improperly apply statistical tests, and draw unfounded conclusions. Take an introductory statistics course or read a book to understand the term statistical inference, power analyses, hypothesis testing, choosing the right statistical test, p-values (including what they mean and that they were never intended to unilaterally signify validity of a result in the way that they are often presented today), correcting for multiple comparisons, Bayesian methods, and other key concepts. If you choose a single book or source, I strongly encourage you to read *Intuitive Biostatistics* by Harvey Motulsky, who created the widely used GraphPad Prism data analysis software. This book is a concise, easily understandable read that will provide you will all of the essentials necessary for most researchers outside advanced computational fields.

ADDITIONAL ACTIVITIES NOT TO BE OVERLOOKED

While the focus of your PhD is and should be research, there are a variety of ancillary components that should not be overlooked because they can significantly augment your progress in research and your overall life balance. Take advantage of academic resources, such as required and elective coursework, individual lectures and seminars, career workshops, mini courses and skill-based bootcamps, internal and external conferences, interest groups, and more. Befriend colleagues; it can be nice to have friends who understand the lifestyle and challenges faced by a scientist, and colleagues in the same, adjacent, or different fields can all provide you with helpful perspectives when you find yourself stumped on troubleshooting an experiment or interpreting a piece of puzzling data.

Seriously consider exploring opportunities to teach, tutor, and/or mentor junior students. This provides opportunities to improve your understanding of key concepts and also contributes to inspiring future scientists and PhD students. Exploring different teaching settings also helps you determine to what extent teaching is involved in your desired career goals. This can range from classroom teaching to hands on laboratory experience or other contexts. We have found that teaching is a highly rewarding and gratifying adjunct to research.

Throughout your academic journey, you will give and receive feedback, and the PhD degree is no exception. "Feedback culture" is becoming more prevalent in academia, in which formal feedback is solicited from all parties at frequent intervals. However, informal feedback given and received on an as-needed basis can be some of the most helpful when taking place in a supportive and nurturing environment. It is important to be able to give and receive constructive feedback in a genuine yet kind and supportive way. There are well-validated methods for accomplishing this, such as the sandwich method in which a negative piece of feedback is sandwiched between two positive pieces of feedback, or the "keep, stop, start" format where an individual is provided with examples of what they should keep, stop, and start doing. We feel that it is best to practice a general compassionate approach, in which the feedback giver expresses appreciation for the receiver's effort, asserts their care for the receiver's best interest, and then delivers the constructive guidance in an honest yet kind way. On the corollary, it is also important to solicit feedback as an important part of your learning and improvement, and to be ready and open to receive both glowing and constructive feedback alike with grace. Ultimately, understanding constructive feedback will allow you to grow and become stronger by incorporating concepts into future efforts, and it also establishes a bond between the feedback giver and receiver.

MILESTONES

The qualifying exam, written thesis dissertation, and oral dissertation defense are unique milestones in the PhD. See Chapter 20 for a discussion of additional potential milestones and crossroads, such as switching advisors, switching research projects, pivoting to a new field, legal considerations, and more.

The qualifying exam typically takes place in years one, two, or three of the PhD after the majority of coursework is completed and a primary research project is identified. The qualifying exam can include components of a written knowledge assessment, oral questioning by a committee, presentation of a research proposal for work to be carried out during the PhD or for unrelated work, or a combination of these features. Passing the qualifying exam is required to advance to PhD candidacy. If a student fails the qualifying exam, they may be required to address deficiencies and retake the exam, or they may be removed from the PhD program – with or without the option to earn a master's degree. Conditionally passing the qualifying exam is often accompanied by specific remediation, which when completed will result in a passing grade. Since the qualifying exam is specific to individual programs/institutions, it is difficult to provide specific guidance. Most important is to consult more advanced PhD candidates who recently completed the qualifying exam in order to understand what strategies allowed them to succeed and whether they identified any pitfalls or made any mistakes that you could avoid. Regardless of the type of qualifying exam administered, it is also important to practice. Assemble an audience of lab members or colleagues and ask them to question you rigorously – the more challenging the better the preparation for the qualifying exam.

The thesis dissertation is a written compilation of all the work performed during your PhD. In the simplest of terms, one can compile thesis chapters based on each manuscript written or paper published during the PhD, with an additional introduction and discussion addressing the chapters as a unit. The dissertation is all-inclusive, long, and detailed – a published work that stands for the immense effort you put in to earn the degree. There are only a few instances where publication of the thesis would be delayed or some work would be omitted from the publication, such as when intellectual property protection prohibits public disclosure of the work for a specified period of time.

The dissertation defense is an opportunity to share a very focused view of your discoveries, placing your work into context of the broader field. The dissertation defense audience is a mix of non-scientist lay people, scientists in other fields, and experts in your field. Thus, the presentation must be carefully crafted to engage each portion of the audience. Often, there will be a public defense session lasting roughly 45–60 minutes more tailored to the lay people and the general scientist audience, with a subsequent closed-door session including only the thesis committee during which more detailed questioning is conducted. As with the qualifying examination, simulation and practice are key to high performance. Assemble practice audiences with your lab group and colleagues in adjacent fields to make sure that the presentation is appropriately tailored to multiple audiences, and demand that they question you as much as possible.

As with any training program, it is helpful to begin identifying potential next steps and applying for subsequent destinations before separating from the prior position. This might involve postdoc training positions, faculty positions, other jobs outside of academia, or entrepreneurial endeavors. Even if the next step is to continue in the same lab as a postdoc or senior scientist, plan in advance for a smoother transition. See Chapter 17 for a dedicated discussion on postdocs, Chapter 18 for faculty positions, and Chapter 19 for non-academic career opportunities.

12 The Secret Sauce

How to Be an Efficient and Effective Research Scientist

Aaron Leconte

In my experience in 20 years in science, efficiency and project management often define who is a "great experimentalist" and who makes groundbreaking discoveries. Even so, it is not actively taught to students and, in fact, most of how laboratory coursework (the first introduction to the lab for most students) is structured can build bad habits that prevent young researchers from being effective and efficient. This book chapter seeks to lay out how to think about project management for young researchers.

I've always been a process-oriented person, and so for many of the things that wound up in the "Secret Sauce," I kind of figured them out during graduate school. I think for many people, this is what graduate school, ultimately, is for, even though nobody actively teaches it to you. But then, during my fourth year of graduate school at Scripps Research Institute, my PhD advisor's own PhD advisor, Dave Collum from Cornell, came to my graduate school to give seminar. As a side talk (distinct from his scientific seminar), in a dark basement, which only added to the mystery of it all, he also gave a presentation only for graduate students that he had been giving to first-year organic chemistry graduate students (including once upon a time, my PhD advisor) for years while at Cornell. I was told by my advisor that I must go to this talk.

This talk, the "Mother Liquor" talk, perfectly distilled all of these untaught rules and ideas of how to be an efficient and effective organic chemist. The talk caused quite a sensation at Scripps at the time, most of the chemistry graduate students attended and, in the last years of my PhD, it was not uncommon to visit a friend in another research lab at Scripps and see a shorthand version of Collum's Mother Liquor talk (which he distributed in little cheat sheets) posted above someone's hood as an ever-present reminder of these principles.

The "Mother Liquor" talk was pretty specific to organic chemistry, but, having been an organic chemist in a past lifetime, I saw how many of these ideals could easily translate much more broadly than just organic chemistry. I was able to get my hands on the slides for Collum's talk, and I always knew that one day, when I had my own lab, this would be a pillar of training.

Since 2012, I've been a PI at a primarily undergraduate institution where my focus is to train the next generation of great experimental scientists. I made it a

DOI: 10.1201/9781003301400-12

point in my first year to translate Collum's talk into a more generalized version (rebranding it as the "Secret Sauce") for an upper-level methods course that I taught. I've been teaching/evangelizing it to undergraduates in my classes and in my research lab ever since, fighting the good fight in the name of Dave Collum to make research more efficient and more impactful.

You may be asking yourself "why is it called the Secret Sauce?" It's a good question. As most of my students will likely tell you, I'm a fan of metaphors that are stretched to the point of nearly breaking (but oddly sort of make sense). So, in this particular strained metaphor, consider a hamburger. The meat is the scientific question, the toppings are the methods, but the "secret sauce" is efficient project management (the topic of this diatribe). All of these elements (meat, toppings, sauce) are necessary for a successful hamburger or project, but, much like most hamburger chains, the Secret Sauce is often mysterious and closely guarded. The goal here is to give every student the recipe.

OVERVIEW

If you learn one thing from the Secret Sauce, it is to be as efficient as possible at answering your research question. You want to run as many experiments as you can, generate as much data as you can from each experiment, and generate as much data in a given time as you can. Most people understand that this is inherently good, but don't know how to do it.

If you learn a second thing from the Secret Sauce, it is that most of the things that allow you to be efficient in a laboratory class (where a lot of training happens for young scientists) are the exact opposite of what makes you efficient as a researcher working on an open-ended problem. Discipline and adherence to habits that make you more efficient in a research setting will make all the difference in your science and will unlock your potential.

YOUR EXPERIMENTS SHOULD FOCUS ON ASKING THE LEAD RESEARCH QUESTION DIRECTLY

Ingredient 1: Progress: Daily, Decisions: Weekly

You should know what hypothesis/research question you are asking with every experiment, and you want to design your experiments with decision-making in mind. A common saying in my lab is "the only failed experiment is an inconclusive one." In other words, if your experiment rejects your hypothesis, then that was productive (even if it is disappointing). A lot of researchers get trapped by "if this works …" experiments. Yes, if it works, that is great! But if you will just try another "if this works" experiment if it fails, you have not made progress if it does not work. Every day, you want to have an experiment running that is asking the key question on your project and will progress the project whether it confirms or rejects the hypothesis that you are querying. This is "Progress: daily."

Oftentimes, researchers can become emotionally invested in an idea, and they can keep trying to make the idea work indefinitely. Create space for a weekly "decision-making" time where you will decide whether you are moving closer to answering your scientific question. Personally, I like to make slides of all my experiments and results once per week to think through what I have been doing in lab. If you are not moving closer to answering your question, then you need to figure out different experiments that can get you closer. If you feel like you are not moving closer to answering the question, and you don't know how else to move closer to answering the question, it's time to look for a new scientific question. This is "Decisions: weekly."

(Note: different research areas can obviously have really different timelines, especially when cell growth or animal usage is factored in. The timelines may change, but the sentiment should be the same: periodically, reflect on "are my experiments moving me closer to answering my research question?" and use that to guide the trajectory of your research.)

INGREDIENT 2: IF YOU ARE NOT AT THE FRONT OF RESEARCH, YOU ARE NOT MAKING PROGRESS

Your time should be focused on pushing your research question forward, not on making the materials that will eventually allow you to ask the exciting, leading-edge question. Making starting materials should be done in background (when you have spare time) and should be done in anticipation of the leading-edge experiment. When you are running an experiment that is your leading-edge question, and you are waiting for the results, be anticipating the next leading-edge question and be thinking about how you can generate materials so that you can immediately progress to the next scientific question rather than waiting to make starting materials. To keep your project moving forward, you must always be at the front of research.

AIM FOR THE MOST EFFICIENT AND MOST MEANINGFUL EXPERIMENTS TO ANSWER THE QUESTION

INGREDIENT 3: DO NOT REINVENT THE WHEEL

Unless the goal of your research question is to develop a method, use methods and procedures that are established and work well. Don't fall into the trap of shiny and exciting new, untested method unless it is actually necessary to ask your question. When you deviate from known procedures or use untested techniques, you are asking two questions: does the new method work AND your research question. If your goal is to get direct answers on your research question (see Ingredient #1), this will slow you down.

To help with this, you should know where to find dependable protocols and what resources have dependable, well-written protocols (bonus points if it has info on troubleshooting). There is a lot of value to be found in relying on classic protocol-focused books like Molecular Cloning or Current Protocols in Molecular Biology/ Current Protocols in Protein Science or well-regarded protocol-focused journals like

Methods in Enzymology. (Note: what resources you use will depend on the field, but find them, particularly when you are starting out in a new field.)

INGREDIENT 4: KNOW WHEN TO CUT CORNERS AND WHAT CORNERS YOU ARE CUTTING

When you modify a procedure, once again, you add a secondary question in addition to asking your research question (now you are asking: did the change in the procedure impact the outcome AND your research question). The first time you run an experiment, you should do it by the book (as mentioned in Ingredient #3).

However, as you grow more comfortable/skilled, you can start cutting corners to make experiments more efficient. Through repeating the experiment many times, you learn what corners can be cut and how to make experiments more efficient. Always be aware of what corners are cut and carefully document them; they are usually the first thing that should be checked when things go wrong.

INGREDIENT 5: WHEN OPTIMIZING: MINIMIZE SCALE WITHIN REASON/MAXIMIZE SHOTS ON GOAL

When trying to identify conditions or piloting a new research question, it is better to run many small reactions rather than one large reaction (it is a more efficient use of time and materials, especially if the starting materials were labor intensive to make). When thinking about "how small," start by thinking about the analytical method that will be the output for the experiment; how much material do you need to feel confident in the results? If you cannot generate enough material to analyze, it is not worthwhile.

The other question to think about is "how many experiments in parallel should be run?" You want to balance quantity of reactions with what is manageable for you (there is no point to setting up 100 reactions if you cannot interpret them/cannot set them up without making mistakes; be self-aware enough to know!). When you are starting out using a method, keep it simple so that you can be confident of the results; as you gain confidence, feel free to add more! A good rule of thumb related to the "how many experiments" question is the "rule of three." Each time you run an experiment, limit scaling up (either in terms of the scale of an individual reaction or in terms of the number of experiments run in parallel) to three times your previous biggest experiment. This will help make sure that you are measuring a manageable quantity/limit the effects of increasing the scale.

DO NOT WASTE TIME REPEATING EXPERIMENTS

INGREDIENT 6: FAILED EXPERIMENTS ARE THE BIGGEST WASTE OF TIME: DO IT RIGHT THE FIRST TIME

There is no bigger time waster than having to repeat experiments, so do it right the first time. ("If you have time to set up a reaction lazily, then you better have time to run it again when it fails.") To prevent repeating experiments, pay attention to what you are doing and document everything carefully as you are doing it. Set up experiments in a way that will not make you want to set it up again if it fails. Setting up experiments quickly and haphazardly does not improve research progress; if you have to rerun it because you didn't get it right, then it was wasted time. Done correctly the first time (even if slowly) is always the fastest way to set up an experiment.

INGREDIENT 7: A GOOD CONTROL(S) WILL PREVENT YOU FROM REPEATING EXPERIMENTS

When you are writing a paper, good controls are needed to rule out alternative hypotheses. When you are running experiments, they can serve a second, equally important role: troubleshooting. Good, unambiguous controls are essential and will significantly speed up your work. You should aim to design controls that show if each step of your procedure is working. This will allow you to be more confident in your results AND will give you essential insight for troubleshooting if it does not work.

INGREDIENT 8: TAKE THE EXTRA FIVE MINUTES TO STORE MATERIALS PROPERLY AND REPRODUCIBLY

If you have to remake something because you misplace it, it is a significant waste of time. Store all of your materials in a systematic and reproducible way. If your lab does not have a policy for how materials are stored, develop your own!

In my lab, all items that get stored in the freezer have a common code of (initials)-(notebook page #).(sample number). So, if I store a plasmid in my freezer box, it might be AML-145.2. This is useful because it connects directly to your lab notebook where you have more room to explain what the sample is. We use a similar method for data storage (*e.g.*, AML-GEL-173.4). Connecting materials and data to your notebook systematically will make it much easier to find when you are writing a manuscript or reviving an idea that you backburnered for awhile. It will also help you if you are not able to finish a project; the next person will be much more able to build on your work if things are stored systematically and clearly.

INGREDIENT 9: PROTECT GROUP AND INDIVIDUAL STOCKS: PREVENT CONTAMINATION AND LIMIT FREEZE-THAW CYCLES OF IMPORTANT MATERIALS

You should know how different types of materials are stored and know best practices for their preservation. This will minimize ruined experiments and time wasted remaking stocks. Put the added time into storing things correctly.

For my lab (we largely work with plasmids and bacteria), that means minimizing contamination risk and freeze/thaw cycles by carefully following rules around handling/storage. Some basic rules in my lab: for all materials that require time and effort to make (or require purchase), minimize contact with pipetters and time not under flame (always take an aliquot and work from that). For group materials (like plasmids created by a student in our group), take only a small amount and then make your own stock for your own routine use. For everyday materials (like antibiotics), create many small aliquots to minimize freeze/thaw cycles.

This will vary a lot depending on the type of research, but putting in effort to know how to store/handle things will pay off in the long run.

INGREDIENT 10: KEEP TRACK OF MATERIALS: SAVE ALIQUOTS FOR ANALYTICAL EXPERIMENTS AND KEEP TRACK OF YIELDS AND QUANTITIES

When experiments fail, you will want to be able to examine why. Keeping track of quantities of materials at different steps, as well as keeping small quantities for analysis later, will help you troubleshoot quickly when things go wrong.

INGREDIENT 11: KEEP A GOOD NOTEBOOK

Your notebook should contain enough detail that someone else in the lab can repeat your experiments without asking you questions. Be sure to leave enough detail so that someone else can find materials that you have created and can find the characterization data used.

USE TIME EFFECTIVELY

INGREDIENT 12: THE HIERARCHY OF TIME

There will always be down time in research. How you spend that down time is one of the most important factors in success. In order of importance, I recommend:

1. Leading-edge research
2. Preparing for more leading-edge research (looking up protocols, preparing materials, etc.)
3. Maintenance tasks (replenish stocks of materials, do a group job, etc.)
4. Experimental play time (try out an idea that you are interested in)
5. Read literature
6. Organize your data (make slides or handouts)

There may be times in your project where it makes sense to prioritize reading literature, or organizing data, or trying out new avenues. So, I would say that the hierarchy is fluid. But at the end of the day, you should make sure that the first three items are taken care of before the others.

INGREDIENT 13: BE ORGANIZED, INTENTIONALLY SET OBJECTIVES WITH A TIMELINE IN MIND, AND EXECUTE THEM

Maintain a calendar, a to-do list, and any other tool you can to stay organized. As you get more proficient, you will have multiple projects and multiple areas of research going at once. Organization will help you make sure all areas of your science continue to progress.

INGREDIENT 14: DON'T LEAVE THINGS FOR TOMORROW THAT ARE EASIER TO DO AT THE END OF THE DAY

Most people (including me), at the end of a long day, are sure that they'll remember what they were doing when they start back again the next day. Most people (including me) usually can't actually remember what they were doing the day before and end up wasting time trying to remember.

All of the things below are MUCH easier to do at the end of the day since it is all freshly on your mind. If you leave it for the next day, it will take twice as long (or longer) and will be worse quality. Before you leave for the night:

1. Make sure your notebook is up to date.
2. Make sure all materials are stored properly.
3. Make a to-do list for the next day.

Before I leave lab for the day, I usually remind myself to do these three things and I do my best to stay disciplined about it. I even have a checklist by my bench to remind myself. It saves lots of time in the long run.

CONCLUDING THOUGHTS

Project management and efficient design and execution of experiments, as detailed here, can often unlock more data and more meaningful data. Of course, there are many other facets of scientific research that can similarly improve your data and act synergistically with the suggestions of the Secret Sauce. For example, lab safety is paramount to successful research (you cannot run experiments if you hurt yourself or others!). Effective use of literature and data analysis are also essential and can lead to more efficient and effective experimentation. Also, sustainability and efficient use of materials can help you run more experiments for the same cost as well as limiting waste. All of these topics are worthy of their own chapters.

Perhaps, more than anything, I hope that the Secret Sauce encourages you to think critically about the process of data collection and encourages you to strive to improve every process of your lab work (Table 12.1). Consider it a starting point to improving your processes!

TABLE 12.1

The Secret Sauce: How to Be an Efficient and Effective Research Scientist (Being an effective researcher is about more than knowing the theory behind experiments. An effective researcher makes the most of their time by setting up as many meaningful experiments as possible. These guidelines will help you transition into being an effective researcher.)

Your experiments should focus on asking the lead research question directly.

Ingredient 1: *Progress: daily, Decisions: weekly*

Ingredient 2: *If you are not at the front of research, you are not making progress*

Aim for the most efficient and most meaningful experiments to answer the question.

Ingredient 3: *Do not reinvent the wheel*

Ingredient 4: *Know when to cut corners and what corners you are cutting*

Ingredient 5: *When optimizing: minimize scale within reason/maximize shots on goal*

Do not waste time repeating experiments.

Ingredient 6: *Failed experiments are the biggest waste of time: do it right the first time*

Ingredient 7: *A good control(s) will prevent you from repeating experiments*

Ingredient 8: *Take the extra five minutes to store materials properly and reproducibly*

Ingredient 9: *Protect group and individual stocks: prevent contamination and limit freeze-thaw cycles of important materials*

Ingredient 10: *Keep track of materials: save aliquots for analytical experiments and keep track of yields and quantities*

Ingredient 11: *Keep a good notebook*

Use time effectively.

Ingredient 12: *The hierarchy of time*

Ingredient 13: *Be organized, set objectives, and execute them*

Ingredient 14: *Don't leave things for tomorrow that are easier to do at the end of the day*

13 Designing a Research Project

Nir Qvit and Samuel J. S. Rubin

DEFINING THE RESEARCH QUESTION

The first step of any scientific research project is to define an interesting question in the context of prior knowledge, which when answered will advance the field or make a meaningful impact through application. Identify a particular research question or objective that you would like to investigate in an area of scientific inquiry that interests you. These may include more basic science research questions (*e.g.*, the mechanism of a theory that explains a natural phenomenon) or more translational research questions (*e.g.*, unmet needs with specific applications in technology or medicine).

To choose an interesting research question, make sure that you are up to date on current knowledge in the field and have sufficient background information to understand the big picture relevance and implications. One approach to gaining this requisite perspective on the current state of knowledge in your area of focus is to conduct a comprehensive literature review of existing research and theories pertaining to open questions. Moreover, it is also important to identify a specific problem, gap, unmet need, controversy, or unanswered question that your research can address. In doing so, clearly state the research question or objective. Furthermore, it is important to make sure you can articulate specifically what the study aims to address and its significance. You should then formulate a set of clear and testable hypotheses or research questions that are aligned with the purpose of your study and will guide your investigation. It is key for you to ask questions that no matter the answer will advance knowledge in the field, even if your hypothesis is incorrect.

STEPS IN DESIGNING A RESEARCH STUDY

After you have formulated your research question and hypotheses, design a few specific sets of experiments that you can use to test your hypotheses. You should develop the methods and procedures to collect and analyze the data. The experimental design should be specific and comprehensive and include detailed information about all steps in the study, including study design, sample size, positive and negative controls, data collection methods and tools, pilot studies to validate and optimize the methods, measured variables and data, and statistical

DOI: 10.1201/9781003301400-13

analyses. Some considerations that should be taken into account when designing the project include:

1. *Experiment design:* It is of the utmost importance to discuss your experimental ideas with colleagues, collaborators, advisors, mentors, and other lab group members. As a general rule, wisdom gained from many minds is always greater than the wisdom gained from a single one. When designing the experiment, consider what will comprise your intervention groups and positive/negative controls. Keep an organized record of all your studies and data collected in a traditional lab notebook or an electronic lab notebook (ELN).

2. *Reagents:* What materials do you need? What are their prices and availability, and are they subject to any regulations?

3. *Workforce and equipment:* How many people are required to conduct the experiment? What expertise is required to conduct the experiment? Are there any specific devices or pieces of equipment required to conduct the experiment? What is the expected duration of the experiment? If longitudinal, how often will you need to physically attend to the experiment and at what hours?

4. *Ethical considerations:* Submit ahead of time any forms or applications for ethical oversight for working with humans (Institutional Review Board, or IRB) or animals (Institutional Animal Care and Use Committee, or IACUC). You will need to address how you plan to handle ethical issues (*e.g.*, obtaining informed consent, protecting participant privacy) and take care of animal welfare by minimizing any potential harm following the principles of the three R's for animal use in research (Replacement, Reduction, and Refinement).

5. *Statistical analysis:* The appropriate use of statistical methods is essential for the conduct of scientific research. When used correctly, statistical tests provide us with means of interpreting the dataset accurately to make appropriate decisions about how to proceed. What is the methodology you will use to conduct your statistical analysis? Is there technical or biological variation in the data that you need to consider? What is the appropriate number of technical and biological replicates for the experiment? What thresholds will you use to inform your interpretation of the data and next steps?

6. *Data collection and analysis:* Determine the method of data collection (*e.g.*, images, matrices, surveys, other observations, etc.) and the variables to be studied as well as how they will be measured. Define how you will process and analyze the data, interpret the results, and answer your research questions using statistical or qualitative methods. Once you have addressed your main question, you may be able to address other questions using the data you collected. Based on your data, is there any way that you could tell a different or additional story?

FUNDING CONSIDERATIONS

For a research project to be successful, it is essential that adequate funding and resources are obtained prior to beginning the work and matching the requirements to complete the project. Investigators can often overcome financial constraints if they carefully consider resource requirements and explore multiple funding options, although this takes time and necessitates creativity and resourcefulness. Options include government grants, foundation grants, industry partnerships, and collaborative academic funds. Commonly, the principal investigator is the one with the responsibility (or headache) of securing the funding for a project; however, it is advisable to become familiar with these opportunities and challenges as early in your career as possible.

The main avenue and most traditional approach to seeking financial resources to support a research project is through grants from governmental agencies, foundations, and research institutes. In order to prepare a grant application, it will be necessary to highlight the significance, the objectives, and the expected results of the project, as well as potential pitfalls and approaches to address them. The chances of receiving funding can be enhanced by collaborating with an experienced grant writer or researcher with a good track record of grant awards. Partnering with companies in industry can be an effective way to obtain additional funding and resources as well. Companies interested in the field may want to become involved in early research projects aligned with their objectives. Collaborating with another academic institution or research organization can also provide opportunities for funding, the sharing of resources, and the sharing of expertise. A partnership or a joint proposal can enhance the credibility of a research project and increase its chances of funding when part of a collaboration or joint effort.

TELLING A STORY WITH DATA

When you have collected and analyzed enough data, think about how the pieces of the puzzle come together to form a story – often an ensemble of studies centered around a single theme. It can be challenging to decide when enough data has been gathered to make a complete report, and rarely will it ever feel complete. The research findings will be presented and should be interpreted as a whole. The results of your study should be compared to the existing literature, the limitations and weaknesses in your study design should be identified, and the implications and significance of your findings should be discussed. A discussion of the implications of the findings should be presented along with the findings themselves. Comparing your study to existing literature can help you identify any weaknesses or limitations of your study and also facilitate the continual design of further experiments. It is critical to communicate your findings to the scientific community in the form of oral presentations, posters, and scientific papers in order to advance the field and update the dynamic scientific record.

SUMMARY

The process of designing a successful research project includes three core steps. In identification of the problem, it is imperative to identify and clearly state the research question, hypotheses, objective, and significance of the study. Choose a challenging problem or question that you can use research to address in some meaningful way that advances the field no matter the result. In experimental design and interpretation, there are many factors to consider while planning experimental procedures (*e.g.*, reagents, workforce, ethical guidelines, statistical analysis, data collection, etc.). Allow yourself plenty of time to plan and prepare before starting the experiment, but do not dillydally. You should ensure clarity, precision, and scientific rigor throughout the design of your study. You should ask for feedback and suggestions from experienced researchers, colleagues, and advisors during the development of your study and periodically as data is collected. Continuously review and refine the design of your research throughout the process of conducting the research, seek feedback from colleagues and mentors, and keep yourself up to date with the latest advances in your field of study by active engagement with the literature. To communicate and disseminate findings, describe the study results as well as their implications in a report that also includes associated contextual information (methods, etc.) to aid in the interpretation of data. In light of the study findings, suggest areas for further research and make recommendations based on the findings. The process of putting together a cohesive and compelling story can be extremely challenging, but at the same time, it is a key part of translating research into meaningful contributions.

14 Obtaining Research Funding

Nir Qvit and Samuel J. S. Rubin

INTRODUCTION

Raising money is one of the most critical, challenging, frustrating, and rewarding responsibilities of a principal investigator (PI). To raise money in most cases, you need to have an impressive track record of publications. Yet to publish you need to do high-quality and high-impact scientific research, which in many cases is also relatively expensive, meaning that you need money (a lot of money).

Research is expensive! First, you need to buy and maintain a variety of devices. While modern laboratory equipment is very costly (in many cases, it is in the range of millions of dollars), unfortunately, maintenance and consumables for these devices are even more costly. Therefore, it is necessary to consider these "extra" costs when buying a device; otherwise, fancy and expensive equipment will be useless. Workforce and effort are often the biggest (and most significant) budget items in lab allocations. There is great benefit from building a diverse team comprised of early career students as well as more experienced trainees such as postdocs, males and females, domestic and foreign individuals, and those with training and expertise across different disciplines. Additional substantial budget items include reagents, consumables, and animals (if the lab uses animals for research). In some places, the "overhead" (the part of a research grant intended to cover indirect costs and allocated to the department, institute, or university) may be very high (depending on the grant agency and university policy). Other minor costs include publication fees, membership in various academic societies, participation in conferences, etc. Therefore, running a research lab involves carefully raising and allocating money. Some PIs claim that their main and most meaningful role is to raise money, and this is what they do most of their time. Although funding procurement takes a significant amount of effort and has become increasingly competitive due to reductions in public funding availability, we hope that other aspects of overseeing scientific research and mentoring trainees can remain central roles of the PI.

For researchers in academia, the main source of funding is government grants. However, funding can sometimes also be obtained from industry partners, philanthropic foundations, research institutes (*e.g.*, Howard Hughes Medical Institute), organizations focused on a particular issue or field (*e.g.*, Bill and Melinda Gates Foundation), private fundraising or individual donations, and even venture capitalists. Each source has its own advantages and disadvantages in terms

DOI: 10.1201/9781003301400-14

of how the funds are obtained, how much money can be obtained at once, and how the funds can be utilized. Here, we focus primarily on government grants since this is the primary source of academic research funding.

Many books, courses, and classes are dedicated to the subject of writing grant proposals. While these applications differ widely across scientific disciplines and grant funding agencies, there are some general guidelines that apply universally. Here, we discuss step by step tips and tricks on how to apply for research grants.

BEFORE YOU START

Allow plenty of time for the grant application process!

Finding the right funding opportunities can be difficult and time-consuming. Since funding opportunities are limited and highly competitive, and writing a grant proposal can be time-consuming, identifying appropriate funding agencies and specific grants or requests for applications (RFAs) is critical. There are many different types of grant agencies for research: government agencies, private corporations, privately funded foundations, professional organizations, scholarships, fellowships, institutional seed funding, prizes, awards, etc. Understanding which funding body funds different types of projects in your field and their suitability to your grant proposal in terms of subject matter supported and resources offered is the first step in the process.

We recommend starting to write grant proposals early and spending ample time reviewing the grant instructions and previously funded successful proposals. Grant applications often have instructions that are over 100 pages. They include information about basic formatting, such as page limits, font types, and sizes, the type of images allowed, citation format, and page margins. All these instructions should be carefully read, as applications that do not comply will not be accepted for review. Therefore, allow yourself several months to prepare your proposal. Preparation time may vary between different scientists based on the time needed to prepare preliminary data, PI experience level, and proposal length or complexity. Plan ahead so that the grant proposal is ready for submission at least one week in advance in case there are last minute changes or unforeseen technical difficulties.

How do you find the appropriate funding agencies for your project? The most effective way to address this question is by dividing it into several small questions:

1. How much funding will be needed for the project? Before you write a grant proposal, one must decide what kind of research will be done. A project cost estimation (budget justification) should be made carefully, ensuring that all budget items meet a funding agency's requirements. Share the draft budget with the designated university official in charge of grant administration for guidance and assistance.
2. From what sources have other investigators at your institution been successful in obtaining funding? Academic research institutions usually have an office responsible for research funding. In many cases, this should be your starting point for identifying funding sources and beginning your research proposal. These offices provide a variety of different services

regarding grant proposals, including identifying and communicating with potential funders, as well as developing proposals. Evaluate the types of programs offered by a funding body and visit their website to study it thoroughly. In many cases, you can and should contact the agency directly (*e.g.*, program officers at the National Institutes of Health) for advice regarding the application process. While most funders will not give specific recommendations, some will review your proposals and give helpful feedback. Many will clearly outline the requirements and steer you in the right direction.

3. After you have identified a funding source, study their application process. Consult with experienced faculty or staff, especially those who have applied for the same or similar funding mechanisms or sit on grant review panels. Review successfully funded grant application to determine what appeals to the funding body. Gather as much inside information as possible.

WRITING THE GRANT PROPOSAL

Several steps are required to compose a coherent grant proposal. The first and perhaps most critical step is to choose a study question, so expect to spend plenty of time formulating your research question. It is usually more challenging to match a funding stream to your project, so we recommend formulating your project comprehensively and accurately first before identifying funding bodies. Then, solidify and adapt the proposal to adhere to funding agency solicitations.

THE SCIENTIFIC QUESTION

A research question must be balanced between having enough novelty, rigor, and relevance to warrant funding, while also not being too ambitious that it becomes implausible or unrealistic to expect it can be accomplished with the proposed timeline and budget. Therefore, considerable effort should be dedicated toward defining a realistic and attractive research question around which to build the proposal.

The scientific question, specific aims, and associated hypotheses to be tested must be stated clearly. Hypotheses should be significant, leading to a substantial advance in knowledge once proven or disproven. Explain how the proposed research will serve the scientific community, patients, and/or society. A strong hypothesis is clear, novel, feasible, testable, and informative regardless of whether it is proven or disproven.

Begin your proposal by providing compelling background information that (1) allows non-experts to understand how your proposed research fits within the field, (2) shows reviewers that you have the appropriate expertise to conduct the research, and (3) articulates to the review panel how the outcomes will advance the field.

Once you have clarified the scientific question (general aim), break it into pieces (specific aims) that test different parts of the hypothesis in complementary ways. These questions (usually three to five) should be broad, linked but not dependent on each other, hypothesis-driven, and not technical. In writing the question and specific aims, consider whether they meet FINER criteria (feasible, interesting, novel, ethical, and relevant).

EXPERIMENTAL DESIGN AND METHODS

The experimental design and methods section should constitute the majority of the grant proposal. Convince the reviewers that the relevant methodology falls within the team's expertise and access to resources. Assure reviewers that the proposed research design and suggested strategies will appropriately address the aims. Since one of the goals of this section is to convince the grant panel that the selected methods offer the means to efficiently interpret the study outcomes, an effective way to address this issue is by drawing appropriate examples from your publications. Additional preliminary data (often not published yet) is another essential part of a research grant proposal and helps establish the likelihood of success for the proposed project. Sometimes this preliminary data will have been generated using funds from a prior grant, institutional discretionary funds, research funding from a senior advisor, or collaborations.

BUDGETING FOR THE RESEARCH GRANT

To ensure that the proposed research is cost-effective and realistic, a funding agency will ask you for a detailed budget and budget justification. Usually, funders provide information about the monetary limit and what research funds can be used for. While this is mainly a technical part, preparing a reasonable and justified budget based on the project needs and funder instructions is critical. This should be done with the assistance of local institution grant administrators who can guide you on how to create the budgets and complete the budget forms according to the resources you propose to use.

THE REVIEW PROCESSES

Think like a reviewer!

Each funding body has a defined process for reviewing submitted applications. In general, the review includes an evaluation of the proposal's scientific and technical merit. It is usually done by scientists from the extramural research community. These researchers are highly busy people, like most of us. Usually, they are assigned to read multiple grant applications in a limited period of time for each cycle. Therefore, reviewers will not be impressed with a sloppily prepared grant application that ignores agency guidelines or a lengthy confusing saga. In addition, reviewer expertise may not be specific to the grant proposal topic. Their time is valuable, and it is paramount to deliver the grant application in a manner that is easy to understand, concise, and informative. Since many reviewers make up their minds very quickly, it is imperative that you make an excellent first impression. Therefore, the abstract should clearly reflect the study objectives and potential outcomes.

The major source of public funding for research in the United States is the National Institutes of Health (NIH). NIH grant applications are scored based on five criteria: (1) significance (if the project aims are achieved, how will the results advance the field and what impact will they have on society?), (2) approach (are the research design and methods reasonable and appropriate to address the specific aims?), (3) investigators (are the investigators experienced in the field and well suited to the project?), (4) innovation (is the project novel and/or will the investigators develop or use novel techniques, and does the proposal challenge current research or strive to shift practice paradigms?), and (5) environment (will the institutional environment and facilities contribute to the probability of success?). While not all funding agencies explicitly ask for these specific criteria, it is helpful to remember them while tailoring your grant proposal.

FELLOWSHIP AND EARLY CAREER FUNDING

Fellowships represent an alternative source of funding for trainees and early career independent researchers who may have less preliminary data or track record than is typically needed for a traditional major grant application. Obtaining a fellowship is also a valuable endeavor for those seeking advancement in their careers. Fellowships provide an opportunity to engage in intensive research, creative high-risk projects, and/or receive specialized training in a field of interest. Fellowships also offer financial support, mentoring, and the chance to pursue focused endeavors in an environment conducive to their success. Financial assistance is provided along with a stipend that is intended to cover living costs, travel expenses, and purchasing some limited research materials, allowing fellows to devote more time to their research projects. Some early career awards come in the form of mentored fellowship grants that aim to facilitate an investigator's transition to independence. Typically, to be considered for a fellowship, candidates undergo a rigorous application process in which they are evaluated on academic qualifications, achievements, and significance of the proposed project.

The development of a fruitful fellowship application is in many ways similar to a successful research proposal. However, one will also need to provide letters of recommendation to support a fellowship application (although some grant applications require similar letters of support from colleagues or collaborators). Recommenders should be selected who are capable of providing a *strong*, insightful, and detailed assessment of your abilities, achievements, and potential in an objective and constructive manner. Ideally, you should choose recommenders who are familiar with the work you have done in academia or in your professional field. Letter writers can provide a good sense of your qualifications and suitability for a fellowship based on their experience and judgment. Ideally, recommenders should be provided with ample time to complete their letters in a well-articulated and positive manner in order to assist in the application process.

In addition to the financial support provided by fellowships, these programs offer several other benefits. In many cases, fellowships provide an opportunity for individuals to connect with like-minded colleagues, experts, and mentors who can provide guidance and valuable connections. Fellowships facilitate interdisciplinary collaboration, intellectual exchange, and exposure to a wide variety of viewpoints, enriching a fellow's research or creative armamentarium. Moreover, it is important to mention that a fellowship can provide an individual with the opportunity to enhance their professional and academic reputation. With this achievement, you may be able to better access opportunities in the future, such as additional funding and career advancement resources.

SUMMARY

Writing a compelling grant application is an integral skill for the independent academic researcher. Mastering this process is crucial for maintaining that independence. A successful grant proposal requires balancing a creative and innovative hypothesis with a testable research plan, which takes plenty of time. A strong proposal will demonstrate three main elements: (1) a clear focus on a well-written hypothesis; (2) a research plan and methods section that is detailed, organized, and logical; and (3) how the research outcome(s) will enhance, change, or expand current knowledge and contribute to the field and society.

The success rate for research grants is as low as 5–10%, so perseverance is also important when applications are rejected. Before submitting a grant proposal, seek feedback from colleagues and individuals with the appropriate expertise. When a grant application is turned down, ask for the reviews, but do not read them on the same day you received the rejection. After at least a day, read the feedback carefully and try to understand why the proposal was not funded. In many cases, reviewer concerns are general, and this process can guide you in other grant applications. Good Luck!

REFERENCES

Ardehali, Hossein. "How to Write a Successful Grant Application and Research Paper." *Circulation Research* 114, no. 8 (April 11, 2014): 1231–1234.

Arthurs, Owen J. "Think It through First: Questions to Consider in Writing a Successful Grant Application." *Pediatric Radiology* 44, no. 12 (December 1, 2014): 1507–1511.

Brownson, Ross C., Graham A. Colditz, Maureen Dobbins, Karen M. Emmons, Jon F. Kerner, Margaret Padek, Enola K. Proctor, and Kurt C. Stange. "Concocting That Magic Elixir: Successful Grant Application Writing in Dissemination and Implementation Research." *Clinical and Translational Science* 8, no. 6 (December 1, 2015): 710–716.

Chung, Kevin C., and Melissa J. Shauver. "Fundamental Principles of Writing a Successful Grant Proposal." *The Journal of Hand Surgery* 33, no. 4 (April 1, 2008): 566–572.

Gemayel, Rita, and Seamus J. Martin. "Writing a Successful Fellowship or Grant Application." *The FEBS Journal* 284, no. 22 (November 1, 2017): 3771–3777.

Gholipour, Ali, Edward Y. Lee, and Simon K. Warfield. "The Anatomy and Art of Writing a Successful Grant Application: A Practical Step-by-Step Approach." *Pediatric Radiology* 44, no. 12 (December 1, 2014): 1512–1517.

Koppelman, Gerard H., and John W. Holloway. "Successful Grant Writing." *Paediatric Respiratory Reviews* 13, no. 1 (March 1, 2012): 63–66.

Neema, S., and L. Chandrashekar. "Research Funding-Why, When, and How?" [In English]. *Indian Dermatol Online Journal* 12, no. 1 (January-February 2021): 134–138.

Pequegnat, Willo, Ellen Stover, and Cheryl Anne Boyce. "How to Write a Successful Research Grant Application." *A Guide for Social and Behavioural Scientists* (2011).

Russell, Stephen W., and David C. Morrison. "The Grant Application Writer's Workbook." Grant Writers' Seminars and Workshops, LLC (2010).

15 Publishing and Presenting Scientific Findings

Nir Qvit and Samuel J. S. Rubin

PUBLISHING SCIENTIFIC ARTICLES

Scientific publication is the process of composing and submitting a research report to a peer-reviewed journal for publication. During peer review, the submitted work should be evaluated by rigorous scientific standards before publication, although political and financial interests can sometimes corrupt this process (more on that another time). An important criterion for selecting a journal for publication is the number of citations that that journal has received. As a measure of the average number of citations per article, an impact factor is used to determine the so-called quality of the journal. To maximize publicity and reputation, researchers try to publish in journals with high-impact factors. Although there are other ways to disseminate scientific findings, the publication of a scientific article in a peer-reviewed journal is still the main method of sharing findings, and academic promotion is built upon impact factor driven publication. One of the most important decisions to be made when publishing your research is choosing the journal to which you submit and hope to have your manuscript published. This decision will make a big difference on the potential impact of your research. Evaluate the scope of the journal, its focus, and its reputation. Also consider publishing options beyond the traditional journal format. For example, an open research platform that allows rapid dissemination makes the research freely available to the public. By choosing a journal before writing or early in the writing process, you will also be able to tailor your work in such a way that it builds on prior research published in that journal and conforms to its formatting requirements. As a result, editors will more easily appreciate the contribution of your paper to the field.

Timing of publication is also a significant factor worth discussion. As you prepare your manuscript, ensure that you have sufficient and substantial results to present. It is possible that you will need to delay the publication if this is not the case. Alternatively, if you are feeling the pressure of competition from other research groups working on similar or overlapping topics, you might want to consider publishing sooner rather than later, since redundant studies are

DOI: 10.1201/9781003301400-15

(unfortunately) rarely published in high-impact journals and priority is generally (although not always) determined by the date on which you submit a manuscript.

As soon as the target journal is selected, researchers should prepare their manuscript in accordance with the journal's guidelines. Describe your work logically and concisely. This is not an easy task to accomplish. To begin, review examples that have already been published by the journal to which you plan to submit your manuscript. The format will vary depending on the journal you choose, the type of paper you intend to write, the field in which you are writing, and the audience. The standard scientific paper format consists of these general sections: the abstract, the introduction, the methods, the results, the discussion, and the conclusion. A scientific paper should also adhere to specified section, word count, and citation style requirements of the given journal.

Most young scientists are not experienced enough to know how to write an article. Some lack the literary expertise to clearly, efficiently, and effectively convey complex scientific information to the target audience. Many writers also suffer from the condition known as "writer's block," which is an overwhelming feeling of being stuck in the writing process, however hard they try. There are two common types of fear when it comes to procrastinating. First is the fear that one is unable to think clearly enough to explain what has been accomplished, and second is the concern that the product will not be "good enough." To overcome these obstacles, begin by creating your figures (and keep in mind a single point you aim to convey with each one) and then tackle the writing, with the understanding that most of the words in your draft will not be included in your final document. As soon as you have something on paper, you can edit it as many times as you like. To get started, you should make an outline for each section and a topic sentence for each paragraph within that section. Then make bullet points for each section with the specific points you wish to convey. The following step is to write the paragraphs in each of the main sections that make up the body of your article. Take plenty of time to read, edit, re-read, seek feedback from others, and tune your manuscript into a presentable product that stands alone.

Submitting your research paper is the next step in publishing. Each journal will have specific submission guidelines and requirements, so check the instructions for authors for your chosen journal. To be successful in submitting your paper to a journal, you should understand its instructions for authors, write an effective cover letter, become familiar with the journal's submission process, and ensure that your research data is properly prepared for submission.

Once submitted, the manuscript is usually screened by an editor to assess appropriateness for potential publication in the journal. If selected, the manuscript then ideally undergoes a thorough evaluation by experts who assess the validity, relevance, and quality of the research. This peer review is an assessment of your research article by independent experts in your field. Reviewers (or referees) will judge the validity, significance, and originality of your work, and hopefully provide constructive criticism and suggestions to improve the work, which in an ideal world are feasible (but this is not always the case). Based on reviewer comments, the authors may perform additional experiments or analysis and revise the manuscript accordingly. If the article is rejected at any point in the process or if the authors elect not to address reviewer comments, then a new journal is identified (sometimes even suggested by the rejecting journal), the manuscript is adapted to meet its requirements, and the submission process is started over again. If and when the manuscript is accepted for publication, it is subjected to final proofing (editing and formatting) by the journal editorial team. Depending on the complexity of the research and the efficiency of the peer-review process, the publication process may take several weeks to months or even years. Finally, the article is published and made available to the scientific community, with or without a cost to authors and readers depending on the journal.

Not all manuscripts are accepted for publication on the first submission. It is a common part of the publishing process for manuscripts to be rejected, and authors should not be discouraged by this inevitable part of the process. For researchers to be able to navigate the publishing process successfully, they must first understand the reasons behind manuscript rejections in order to learn how to deal with rejections when they occur.

It is common for manuscripts to be rejected because they are perceived as not being "original enough" or having little significance. Editors and reviewers of journals place emphasis on publishing studies that contribute to the field in a positive way. As a rule of thumb, a manuscript will be rejected if it does not present novel insights, does not address a clear research gap, is lacking in relevance, or sometimes if it challenges the prevailing dogma too much. Therefore, it is very important that authors conduct an in-depth literature review to identify existing knowledge gaps that need to be addressed by the research they propose. Poor research design or methodology can also result in a manuscript being rejected. Manuscripts that lack a clear research question, have an experimental design that is flawed, or present weak methods of data analysis are likely to be rejected by editors. Authors should ensure that their research design is rigorous, the methodology is appropriate, and the data analysis is accurate. The quality of the writing and the structure of the manuscript play an important role in the decision to reject the manuscript as well. It is more likely that manuscripts that contain errors in grammar, are unclear in their language, or have inconsistent formatting will be rejected. Authors should pay particular attention to clarity and coherence of the text while writing, adhere to the journal's formatting and structure guidelines, and ensure proper grammar and punctuation throughout the manuscript.

Although receiving a rejection can be disheartening, it is imperative that authors handle rejections in a professional and positive manner. You should not dwell on any disappointments or frustrations you may experience as a result of a manuscript rejection. Taking a break from the paper is beneficial and you should engage in activities that will help you relax and regain a sense of perspective until you can return to it later on. Seek feedback from the editor or the reviewers of your article in order to understand why it was rejected. In order to gain insight into your rejection, you may want to discuss it with colleagues, mentors, or writing groups. Collaboration is one of the most effective ways for you to improve the quality of your research and increase the chances of success. Make sure that you are able to utilize feedback constructively so that your manuscript can be improved. If your paper is rejected with an opportunity to revise and resubmit it, make sure that you carefully consider the reviewers' comments and make any necessary revisions to the paper wherever possible or explain why this was not accomplished. If you resubmit to the same journal, make sure that the changes you have made are clearly described in your resubmission. Consider submitting your manuscript to other journals (one at a time) in the same field to increase the chances of publication as well. Depending on which journal you are submitting your manuscript to, you may need to adapt it to their guidelines and requirements. Perseverance, professionalism, and a commitment to continuous improvement are all essential ingredients for navigating the scientific publishing landscape effectively and successfully.

Corrections or retractions of published papers are best avoided, as this represents a major red flag in an author's record, especially in the case of retraction. If a publication requires a minor correction (*e.g.*, an author name spelling, a particular figure title that clarifies the meaning, etc.), this can usually be accomplished without significant negative repercussions on one's career. However, corrections are still best avoided, as there can be negative effects, such as reduced likelihood of publishing again in the same journal and developing a reputation of sloppiness. Retraction of publications, on the other hand, can make a significant negative impact on one's career and their subsequent ability to publish or obtain grant funding. The most common causes for retraction include plagiarism, inappropriate use of statistics (intentional or unintentional), and manipulation or fabrication of data. The best way to avoid retractions (and corrections) is to avoid these malfeasant practices, pay attention to every detail in the submitted manuscript, and double, triple, quadruple, and third-party check every calculation and data analysis process. It is never worth rushing a publication or drawing a sexier conclusion that the data supports if it risks necessitating a future correction or retraction: it is simply not worth it, and it is an ethical assault on the scientific record and often a misuse of public funds.

SCIENTIFIC PRESENTATIONS

Whether just beginning their careers or steeped in experience, students are required to deliver scientific presentations regularly. This is also a major duty of principal investigators, who are responsible for presentation of research findings to diverse audiences for the purpose of teaching, collaboration, and fundraising. There are many more lectures or talks given than papers published. Therefore, the presentation of scientific findings at courses, lectures, meetings, and conferences is another important channel for researchers to communicate their research results. In particular, conferences allow scientists from different locations or from different disciplines to exchange ideas and discuss research findings. Presentations at conferences may take the form of oral presentations, posters, or workshops. During the presentation, researchers highlight key aspects of their study, present their findings, and engage in a collaborative discussion with their colleagues and peers. The majority of conferences are held in person, allowing for the exchange of ideas, feedback, and collaboration between participants.

ORAL TALKS

At most conferences, there is a good chance that you will only have 10 to 20 minutes to present an oral presentation during your session. Therefore, you are not able to present all of your data. Your presentation should emphasize the importance of your findings in a compelling manner. Make sure you give a brief (approximately 60 second) overview of the literature so that people in and out of the field can follow along with your "story," and highlight your original contribution to the field. Consider explaining to your audience that you are only going to present a portion of your paper and that you would be happy to answer any questions that they might have after the presentation in the question-and-answer session that follows. Remember that your audience members are not readers but listeners, and you should keep this in mind when delivering your message. Whenever you present results, try to engage your audience with the context and impact of your findings. Start your presentation with an introduction, then explain what you will be discussing (*e.g.*, what the issue is), tell the story, and then conclude by summarizing what you have accomplished (*e.g.*, what you did to solve the issue). Ensure that you do not get lost in the details of your findings and that you are able to summarize your findings in a few concise points. The first time you present your research, it can feel quite intimidating. However, it is important to remember that no one knows your work better than you. Moreover, you should practice, practice, and practice again. For some students learning the first two slides by heart is very helpful, yet we do not recommend memorization of the presentation verbatim.

POSTER PRESENTATIONS

A poster presentation may be requested or offered as part of a symposium or conference. The purpose of a poster presentation is to facilitate one-on-one and small group conversations between the presenter and visitors to the poster. The poster is usually displayed in a large hall, and the presenter stands by the poster to provide a short presentation and engage in discussion about the research. Poster presentations are similar to oral presentations since they present a limited summary of the main points often found in the paper. But in contrast to an oral presentation, it is easier for the smaller poster audience to interrupt you, ask questions, and grill you regarding your research without putting much effort into it. Nevertheless, scientific poster presentations provide an excellent opportunity to receive feedback and interact with scientists. Having others show interest in your research is a great confidence booster as well.

There is a lot of work involved in preparing a good poster, but it is well worth the time and effort to showcase your research in the best possible light and engage with the audience. The poster should follow a coherent visual path, telling a story efficiently. Each section should follow clearly from the previous one. Your poster should be designed so that anyone reading it (even if you are not around, as they are often left on display outside of the poster session hours) can understand your work. In the same way as any other publication, you should include sections in your poster that summarize the background and rationale, the methodology, the results, and the implications of what you have done. This time, however, instead of relying on text to tell the story, it can be more effective to use figures that illustrate your findings. Using bullet points to summarize sections and highlight key points can provide interpretations and shorten your text even further. A poster with fewer words will better attract the attention of the audience because it looks more inviting. Use graphics, tables, and photographs to present your data. For a poster presentation to succeed, it must be well-designed to catch the audience's attention. Posters should also include professional contact information and logos of your institution and any funding organizations in order to take advantage of the networking opportunity afforded by poster presentations.

After you have created a poster, it is time to prepare for its presentation. It is not necessary to prepare a perfect speech for the poster session since it is less formal than an oral talk or lecture. Just make sure that you can talk about all materials displayed on the poster, answer questions, and refer to a few key talking points for audience members who prefer that you guide the conversation. Take advantage of the poster session to engage in fruitful discussions regarding your data. In order to have more interesting discussions, allow visitors to ask questions about the aspects of your research that are most interesting to them. Since many attendees wish to visit several posters during the session, make sure you are able to summarize your work within a few minutes in order to facilitate their interactions with your poster in an efficient manner.

SUMMARY

There are many advantages to the publication and presentation of scientific findings. This process contributes a great deal to the development of a collective scientific knowledge base. Whenever scientists share their research, they are enabling others to build on their work, which in turn fosters scientific advancement as a whole. Science progresses as a result of the accumulation and dissemination of knowledge, which is a continuous process for which we all share responsibility.

Additionally, publication and presentation of research findings allow scientists to establish their expertise in their field and their credibility within the scientific community. The publication of research articles in reputable journals enhances the reputation of researchers and builds their academic careers. In a similar manner, presenting at conferences is a demonstration of active involvement in the scientific community and facilitates networking opportunities as well.

Through sharing findings with the scientific community, researchers also receive feedback, critique, and/or validation of their work. Constructive criticism received during presentations and conference discussions helps investigators refine their work and address any limitations or potential errors. Peer review assesses the research, ensuring accuracy, robustness, and relevance.

In summary, publishing and presenting scientific findings are essential components of the research process. It is through these activities that researchers are able to share their work, contribute to the scientific community, establish their expertise, and receive feedback. Scientists promote advancement of knowledge and foster collaboration by actively participating in dissemination of research findings.

REFERENCES

Ascheron, Claus. *Scientific Publishing and Presentation: A Practical Guide with Advice on Doctoral Studies and Career Planning*. Berlin, Germany: Springer Nature, 2023.

Fah, Tong Seng, and Aznida Firzah Aziz. "How to Present Research Data?" [In English]. *Malays Fam Physician* 1, no. 2–3 (2006): 82–85.

Quinn, Charles T., and A. John Rush. "Writing and Publishing Your Research Findings." [In English]. *Journal of Investigative Medicine* 57, no. 5 (June 2009): 634.

Silvia, Paul J., and Katherine N. Cotter. "Presenting and Publishing Your Research." In *Researching Daily Life: A Guide to Experience Sampling and Daily Diary Methods*, 127–138. Washington, DC: American Psychological Association, 2021.

Soranno, Patricia A. "Six Simple Steps to Share Your Data When Publishing Research Articles." *Limnology and Oceanography Bulletin* 28, no. 2 (May 1, 2019): 41–44.

16 Science in a Pandemic

Samuel J. S. Rubin and Nir Qvit

The COVID-19 pandemic caused by the SARS-CoV-2 virus caused millions to lose their lives or livelihood and challenged every industry; science and research were not spared. Scientists grappled with new obstacles, and the conduct of research has in many ways been permanently altered by adaptations made early on in during the pandemic. While the world faced many pandemics throughout history, most recently those caused by influenza viruses, we saw spread with SARS-CoV-2 faster than ever before – in part due to intrinsic viral properties and also due to globalization. In many ways, modern science revolutionized our ability to press on, while in other ways major issues pertaining to the relationships between science and society emerged with force.

Like many fields, science adapted to remote learning environments, virtual interactions, distanced siloed work, and more. To an extent, much of this continues to shape the ongoing work today. What's more, public perception of science and attitudes toward research were drastically impacted by the COVID-19 pandemic due to extreme politicization experienced during the time period. Politics permeated scientific, medical, and public health decisions seemingly more prominently than before, although perhaps this is an illusion caused by media coverage and the rise of social media. Prominent appointed and self-proclaimed representatives of the scientific and medical communities made too many missteps and demonstrated poor judgment time and time again, substantially tarnishing reputations of these fields for many. Simultaneously, politicization of medicine and science during this period encouraged many to reflexively endorse concepts and ideology, majorly weakening ability for many to think critically and skeptically when appropriate.

COMPUTATIONAL VERSUS "WET LAB" SCIENCE

A growing interest in computational versus traditional "wet lab" science predated the COVID-19 pandemic. Several of the major motivations for interest in computational research over the past two decades include advances in experimental technologies generating larger datasets paired with innovations in computational approaches capable of analyzing those "big data" sets in ways that presented new and exciting opportunities, as well as benefits in lifestyle for the computational scientist, namely predictable hours and primarily desk work. Thus, computational expertise has become increasingly marketable and value-enhancing across scientific research. During the pandemic, computational studies soared. While scientists were

DOI: 10.1201/9781003301400-16

stuck in place at times without access to their labs, many learned or expanded upon their computational skills and accessed existing or publicly available datasets. This explosion of re-analysis and meta-analysis produced a large number of "dry" studies and also fueled an already growing arena.

VIRTUAL AND REMOTE WORK

With widespread adoptions of advanced communications and networking software, virtual and remote work prevailed early in the pandemic and continues to be widely used in many settings. Lectures and meetings rapidly transitioned to online video formats, and remote exam solutions even provided for virtual proctoring. Conferences, networking events, and interviews also transitioned to online formats, and even since the return to in person activities many of these events have retained hybrid options. Accustomed to conducting work from the comfort of home or another remote location, many still prefer to engage with activities remotely, which will no doubt have lasting impacts across fields. Many are concerned that this will hamper the collaborative nature of scientific investigation. For the time being, much wet lab work still requires in person efforts, although advances in automated lab equipment and cloud-based laboratories may alleviate more of these responsibilities in the years to come.

IMPACTS ON RESEARCH

The conduct and content of research evolved drastically during the COVID-19 pandemic. Lab research was affected by supply chain problems and shortages of vital materials. Many labs redirected their efforts toward COVID-19 regardless of their prior fields of expertise. Translational studies and clinical trials were decimated early on, especially for indications in cancer where many patients were vulnerably immune suppressed (Fox et al., 2021). There was also a general reduction in non-COVID-19 clinical trials (Lasch et al., 2022), stalling many therapeutic candidates in the pipeline.

There was particular strain on junior researchers. Initiation of new projects declined (Gao et al., 2021). Obtaining funding and career advancement became even more challenging ("A Conversation on the Effects of the COVID-19 Pandemic on Academic Careers with Junior Researchers," 2021; "A Conversation on the Effects of the COVID-19 Pandemic on Junior Researchers' Careers with Funders and University Leaders," 2021). Perhaps worst affected were female scientists, bench scientists, and scientists with young children (Myers et al., 2020), as well as minoritized and marginalized early-career scholars (Douglas et al., 2022).

Extraordinary changes also shaped publishing and publications, causing huge ramifications throughout research and science given the use of publications as a major metric for advancement in academia. COVID-19 represented the first pandemic that the modern scientific publishing industry faced, with 1.5 million articles added in 2020 alone, representing the single largest yearly increase to date (Clark, 2023). Submissions increased and turnaround times decreased. There were changes to the ways that papers are produced and vetted. An early effort was made to promote open science with open access to scientific material, and many authors posted their non-peer-reviewed submissions on preprint servers widely accessible to the public. Although posting of preprints has become far less common since the early pandemic, there is data supporting the robustness of evidence reported in preprints, arguing for their continued use despite lack of regulation (Nelson et al., 2022). When understanding of COVID-19 was minimal, editors accepted less substance for publication rather than requiring more time for revisions and decisions to be made. Many saw opportunity in the twisted industry of scientific publishing that thrives on competitiveness. These industry-wide accelerations resulted in a flood of junk literature, fraud, and high-profile retractions, diluting remaining substantive studies. Trends in the scientific publishing industry also led to covidization, the domination of COVID-19-related papers over other topics (Ioannidis et al., 2022). As a result, important topics were neglected and citations were hoarded, significantly distorting journal impact factors, another metric utilized in academic promotion. This climate of opportunism reflected an already perverse academic publishing system resting on a "publish or perish" mentality in academia. While publication ethics may in part be a casualty of the COVID-19 pandemic, it is also to be expected that some inaccurate and/or incomplete information arises during an ongoing emergency, compromising the ability to conduct rigorous studies (Tripathy, 2021). For a more detailed discussion of scientific publishing, see Chapter 15.

Overall, the COVID-19 pandemic permanently shaped the way science is conduced and communicated. Time will tell and many will debate the effects of these changes. For better or worse, scale and speed of research have increased, with more remote work and virtual collaborations.

REFERENCES

"A Conversation on the Effects of the COVID-19 Pandemic on Academic Careers with Junior Researchers." *Nature Communications* 12, no. 2097 (2021). 10.1038/s41467-021-22039-w

"A Conversation on the Effects of the COVID-19 Pandemic on Junior Researchers' Careers with Funders and University Leaders." *Nature Communications* 12, no. 2096 (2021). 10.1038/s41467-021-22040-3

Clark, J. "How COVID-19 Bolstered an Already Perverse Publishing System." *BMJ* 380 (2023): 689. 10.1136/bmj.p689

Douglas, H. M., I. H. Settles, E. A. Cech, G. M. Montgomery, L. R. Nadolsky, A. K. Hawkins, G. Ma, T. M. Davis, K. C. Elliott, and K. S. Cheruvelil. "Disproportionate Impacts of COVID-19 on Marginalized and Minoritized Early-Career Academic Scientists." *PLoS One* 17, no. 9 (2022): e0274278. 10.1371/journal.pone.0274278

Fox, L., K. Beyer, E. Rammant, E. Morcom, M. Van Hemelrijck, R. Sullivan, V. Vanderpuye, D. Lombe, A. T. Tsunoda, T. Kutluk, N. Bhoo-Pathy, S. C. Pramesh, A. Yusuf, C. M. Booth, O. Shamieh, S. Siesling, and D. Mukherji. "Impact of the COVID-19 Pandemic on Cancer Researchers in 2020: A Qualitative Study of Events to Inform Mitigation Strategies." *Frontiers in Public Health* 9 (2021): 741223. 10.3389/fpubh.2021.741223

Gao, J., Y. Yin, K. R. Myers, K. R. Lakhani, and D. Wang. "Potentially Long-Lasting Effects of the Pandemic on Scientists. *Nature Communications* 12, no. 1 (2021): 6188. 10.1038/s41467-021-26428-z

Ioannidis, J. P. A., E. Bendavid, M. Salholz-Hillel, K. W. Boyack, and J. Baas. "Massive Covidization of Research Citations and the Citation Elite." *Proceedings of the National Academy of Sciences of the United States of America* 119, no. 28 (2022): e2204074119. 10.1073/pnas.2204074119

Lasch, F., E.-E. Psarelli, R. Herold, A. Mattsson, L. Guizzaro, F. Pétavy, and A. Schiel. "The Impact of COVID-19 on the Initiation of Clinical Trials in Europe and the United States." *Clinical Pharmacology and Therapeutics* 111, no. 5 (2022): 1093–1102. 10.1002/cpt.2534

Myers, K. R., W. Y. Tham, Y. Yin, N. Cohodes, J. G. Thursby, M. C. Thursby, P. Schiffer, J. T. Walsh, K. R. Lakhani, and D. Wang. "Unequal Effects of the COVID-19 Pandemic on Scientists." *Nature Human Behaviour* 4, no. 9 (2020). 10.1038/s41562-020-0921-y

Nelson, L., H. Ye, A. Schwenn, S. Lee, S. Arabi, and B. I. Hutchins. "Robustness of Evidence Reported in Preprints during Peer Review." *The Lancet Global Health* 10, no. 11 (2022): e1684–e1687. 10.1016/S2214-109X(22)00368-0

Tripathy, J. P. "Is Publication Ethics Becoming a Casualty of Covid-19?" *Indian Journal of Medical Ethics* VI, no. 1 (2021): 1–3. 10.20529/IJME.2020.092

17 Excelling in Your Postdoc

Nir Qvit and Samuel J. S. Rubin

INTRODUCTION

Postdoctoral positions are one of the most popular routes for graduate students to acquire additional experience to achieve advanced career goals. Postdoctoral positions are often necessary although not technically required for academic careers as a principal investigator (PI) because this advanced training period provides opportunities for further professional development and growth into a position of independence. As a postdoctoral researcher, you can benefit from having a clear career plan, demonstrating strong research skills, generating valuable preliminary data required for grant applications, cultivating professional relationships, and participating in professional development activities to ensure a successful post-doctoral experience and transition to subsequent employment.

While the benefits of being a postdoctoral researcher are many, it is important to acknowledge the challenges and uncertainties that come with this phase in the training. A major challenge is the intense competition for limited faculty positions that postdocs often seek, which is a significant obstacle to exiting the position. There is a high level of competition for faculty positions in the present academic job market due to the fact that the amount of late stage postdocs outnumber open faculty positions by a large margin. Therefore, postdoctoral researchers must distinguish themselves by performing high-quality research, publishing in reputable journals, and developing a robust research agenda. With this approach, it is difficult but feasible for postdocs to enhance their visibility within the academic community. A further challenge for postdoctoral fellows is limited job security, as well as the temporary nature of their positions. Postdoctoral researchers should be proactive in planning their career trajectories, exploring alternative career paths if desired, and seeking advice from mentors and career development offices in order to realize their full potential.

ACADEMIC CAREERS

A PI is the most traditional career path following an academic postdoc. An academic PI oversees a lab or research group at an educational institution and plays an active role in research, teaching, and mentoring and is usually responsible for obtaining funding to support their research program. In preparation to become a PI, a postdoc can further enhance their credentials in a given field by developing deeper

DOI: 10.1201/9781003301400-17

or new expertise in a particular technique, expanding their professional network, and nurturing relationships. A postdoctoral researcher should be involved in cutting-edge research projects, collaborate with leading investigators, and work to improve their teaching abilities as well. Despite the fact that the path to a faculty position can be challenging and competitive, one of the main benefits of a postdoctoral position is that it provides the opportunity to gain essential experience, refine one's research goals, and develop an independent research program, all of which are prerequisites to landing an academic PI job.

A postdoctoral position offers many unique advantages to those who wish to pursue an academic career. As a postdoctoral fellow, you will gain a deeper understanding of the intricacies and complexities of a particular research area at a higher level than a graduate student. This includes acquiring cutting-edge research techniques and methodologies that further your research toolkit, as well as navigating more funding responsibilities and academic politics. Postdocs have greater exposure than graduate students to other science-adjacent aspects of research, such as political relationships in field of interest, grant applications, and publishing. As a result, postdocs are poised to make significant contributions to their field and become recognized as rising experts. Postdocs may also receive opportunities to collaborate with renowned researchers in the field. It is through these collaborations that knowledge can be exchanged, interdisciplinary approaches can be fostered, unusual opportunities for funding and facilities can be found, and future jobs can be obtained. Thus, the postdoc allows you to expand your professional network in addition to your technical expertise.

In the course of a postdoctoral position, individuals may have the opportunity to assist in teaching undergraduate or graduate courses, mentor junior trainees, supervise student research projects, or create their own courses. Hands-on experiences like this allow refinement of teaching methodologies, improvement of communication skills, and a deeper understanding of the learning process. By engaging with students, postdocs are able to mentor and inspire the next generation of scholars, fostering a sense of fulfillment and contributing to the academic community. Academic institutions require faculty members to excel in a wide range of disciplines, including cutting-edge research (as evidenced by the publication of high-impact papers and their selection for prestigious fellowships or grants), professional teaching, and effective mentoring, which reflects the diverse responsibilities and roles that faculty members play in academia. A postdoc training position allows the future academic PI to hone their diverse skillset and optimize their candidacy for job offerings.

INDUSTRY CAREERS

A significant number of postdocs choose subsequent careers outside of academia. In addition to postdoc positions in academic settings, there are postdocs offered in industry and government as well, which provide first-hand exposure to those environments. A postdoctoral career in industry can provide excellent opportunities for researchers to bridge the gap between academia and industry. This career path will provide the opportunity to conduct postdoctoral research in a corporate setting.

You will collaborate with industry professionals, participate in applied research projects, and contribute to the development of innovative products or technologies. An industry postdoctoral position that is successful may lead to a permanent position. Companies value the expertise and experience gained by postdoctoral researchers, especially in industry settings. Consequently, postdocs from both academia and within industry are attractive candidates for commercial research and development positions or other jobs requiring specialized knowledge in industry.

Following postdoctoral training, whether at an academic institution or in industry, you may choose to pursue a career in industry. Many industry jobs are available for scientists, and some companies conduct research in a very similar way to the academic community. For example, some pharmaceutical companies have research groups that conduct research under the supervision of a PI in a similar manner to academic research groups, although funding is usually determined by the company and their commercial priorities. Industry positions can give individuals the opportunity to further develop their skillset, network with other professionals, and tackle "real-world challenges," all while emphasizing teamwork, which is particularly appealing to some individuals.

KEY STRATEGIES TO MAXIMIZE YOUR POSTDOC

The postdoctoral position is one of the most important stages in the development of an academic or research career. It is intended to serve as a bridge between the completion of a doctoral degree and the securing of a permanent position after completing the program. With a successful postdoctoral experience, you can enhance your research skills and position yourself to pursue a variety of different career opportunities. These skills include problem-solving skills, critical thinking skills, communication skills, project management skills, network-building skills, and the ability to lead and work as part of a team including members with a variety of different backgrounds. You will have a better chance of attaining your dream job if you develop advanced leadership, communication, and project management expertise to stand out from the crowd. We outline key strategies below that can assist you in maximizing your success as a postdoctoral researcher.

A proactive approach to your postdoctoral experience is the key to your success, as well as effective time management, collaboration with others, networking, and continuing to enhance your professional development throughout your tenure. Investing in professional development workshops or courses and engaging in collaborations will help you maximize the benefits of your postdoctoral experience. By doing so, you will be able to lay the foundation for a successful academic or industry research career built upon a strong network.

At the start of your postdoctoral research, you should set specific, attainable research objectives in order to achieve your goals. Discuss your objectives with your mentor and develop a detailed plan of action that aligns with the lab's interests but also allows you to create a niche that you can utilize if and when you start your own independent research program (you should have enough overlap with your postdoc lab such that you share interests and required resources but also enough separation such that you can develop an independent research program in which you are not a competitor or threat to your PI). Organize your time by creating a long-term schedule, setting priorities, and breaking down your work into smaller steps so you can manage your time as efficiently as possible. Delegate as much of the tangible work as possible to other members of your research team and collaborators; this is your time to practice functioning as a pseudo PI without all the intrinsic pressures of being a fully independent PI. Make sure you monitor your progress and make adjustments as necessary to ensure that your projects are advanced in a timely manner and with maximum productivity.

Share your findings. Your research should be presented at conferences, you should collaborate with your mentor and peers, and you should submit papers to reputable journals. As a result of your presentations and publications, you will be more visible to the scientific community and prospective employers. Participate in group discussions and engage in dialog with lab members to foster an environment of collaborative learning and inquiry. Participate in networking and educational events (*e.g.*, laboratory meetings, seminars, and journal clubs) and present your research on various platforms (*e.g.*, posters, oral presentations, and abstracts) to exchange ideas and receive valuable feedback on your work. Take advantage of opportunities to collaborate with colleagues on projects and share your expertise with them. Maintain meaningful relationships with individuals, exchange contact information, and develop a longitudinal relationship with them. Attend conferences and symposia in your field in order to keep up-to-date on the latest research that is being conducted and maintain visibility in the field.

Take advantage of every mentoring opportunity that comes your way, both as a mentee and as a mentor. Consult with senior researchers and provide support to junior members of the lab as much as possible. Mentorship is an effective way to enhance your leadership skills and contribute to the development of a positive lab culture and scientific community overall. You might also learn something unexpected in the process.

Establish your professional online presence by utilizing platforms such as LinkedIn, ResearchGate, ORCID, institutional webpages, and/or a personal website. Regularly update your profile, share your research accomplishments online, and keep in touch with colleagues and potential collaborators so that you can exchange ideas and resources. Your network can be further expanded by participating in online discussions and joining relevant professional groups that in your field.

Make sure that you take advantage of the training programs, workshops, and seminars offered by your institution or professional society to get the most out of your education. Seek experience in grant writing, scientific communication, leadership, and project management to become a more competent manager. A longitudinal commitment to learning will no doubt enhance your CV.

Discover opportunities for developing transferrable skills, such as those related to data analysis, programming, or teaching, as well as those that are specific to your field. Having these expertise can give you a chance to broaden your career options and make yourself more marketable to potential employers.

In the course of your postdoctoral career, it is important to engage in career planning activities as early as possible. Explore various career paths available to you and establish your career goals and career objectives. Participate in career development workshops, take advantage of mentorship opportunities, and develop a network with professionals from a variety of fields by connecting with them at networking and educational events. With a broad network, you will be able to gain early insight into your future career options and make informed decisions regarding your career path.

There is a wide variation in the length of postdoctoral fellowships offered in STEM fields. The complexity and size of a research project play an important role in determining the duration of postdoctoral positions. Since postdocs are expected to demonstrate outstanding performance to be considered for faculty or other job positions, they often seek projects that are high risk and high reward in nature. These projects are often quite complex, and they can take a long time to accomplish substantial results. A postdoctoral fellowship's duration is very much determined also by the availability of funding for the research and the duration of the research program itself. As a result of limited funding, researchers may have to complete their projects within a shorter period of time. Those who have secured their own funding may be employed as postdoctoral researchers for longer periods. The length of the postdoctoral fellowship may vary depending on the career aspirations of postdoctoral fellows as well. While some researchers aim to become academic faculty, others may pursue careers outside of academia (*e.g.*, in industry) as a means of transitioning into a successful career earlier.

SUMMARY

A postdoctoral fellowship can represent a transformative stage in a scientific research career, serving as a bridge between completing your doctoral degree and securing a permanent position. Excelling in your postdoc is essential for enhancing your research skills, expanding your network, and obtaining subsequent jobs. Focusing on key aspects of research productivity, collaboration, networking, and professional development will allow you to make the most of your postdoctoral experience and position yourself well for subsequent employment opportunities.

REFERENCES

Compton-Daw, Emma. "So You're a Postdoc. What Next?" 2018.
Forand, Nicholas R., and Allison J. Applebaum. "Demystifying the Postdoctoral Experience: A Guide for Applicants." *The Behavior Therapist* 34, no. 5 (2011): 80–86.
Huang, K. L. "Ten Simple Rules for Landing on the Right Job after Your PhD or Postdoc." [In English]. *PLoS Computational Biology* 16, no. 4 (April 2020): e1007723.

18 Starting a Laboratory or Research Group

Nir Qvit and Samuel J. S. Rubin

ACADEMIC, INDUSTRY, AND GOVERNMENT ENVIRONMENTS

Establishing your own laboratory is an exciting endeavor that requires careful consideration of a number of factors, including the setting of the laboratory. Academia, industry, and government represent distinct environments with unique opportunities and challenges for establishing a new laboratory. In this section, we examine the characteristics of each setting and discuss the factors to consider before making such a decision. Ultimately, we focus the majority of our discussion on the academic lab, although many of the concepts and considerations are transferrable to other settings.

Academic careers provide scientists with the opportunity to pursue their passion for research, engage in teaching, and contribute to the knowledge base of their respective fields. The academic environment is intellectually stimulating, with a strong emphasis on critical thinking and scholarly pursuits. Academic scientists choose their research topics, apply for funding, and often collaborate with colleagues from a variety of disciplines. One of the primary advantages of an academic career is tenure and intellectual freedom. Tenure provides job security and the academic freedom to pursue questions of interest, allows scientists to establish long-term research programs, and supports mentoring the next generation of researchers. Being in academia also presents challenges. Faculty are often responsible for obtaining their own funding. The path to tenure is highly competitive and requires dedication and perseverance. Academic positions also often come with heavy teaching and administrative responsibilities, leaving limited time for research.

Mentoring and teaching are integral components of academic environments, as they play an important role in shaping the next generation of scientists and researchers. Mentoring and teaching also provide opportunities for professional and personal development. As mentors, we gain satisfaction from witnessing the progress and achievements of our mentees who contribute to the advancement of science as a whole. However, mentoring also poses challenges, such as adapting to diverse learning styles and individual needs. Moreover, the academic process requires a significant amount of time and energy from mentors. For early-career scientists, balancing research and teaching responsibilities can be demanding.

Pursuing a research niche in science allows individuals to focus their expertise on a particular area of interest and become pioneers in their fields of interest. Scientific knowledge continues to expand and evolve, and these niche careers will

DOI: 10.1201/9781003301400-18

play an increasing role in pushing forward the boundaries of scientific under-standing. Furthermore, they will address the complex challenges facing our world today. As such, these niche careers allow researchers to apply for specific funding mechanisms tailored toward their interests. In embracing a research niche, scientists can make significant contributions to society while pursuing their passions and curiosity in scientific research.

The industry sector includes a broad range of organizations, including pharma-ceutical, biotechnology, and technology companies. In these settings, scientists often focus on translating research findings into practical applications, developing new products, and improving existing technologies. Scientists in industry often work in multidisciplinary teams, collaborating with engineers, business professionals, and other experts to achieve common company-determined goals. A significant benefit of working in industry is the possibility of receiving a substantial salary without the need to apply for funding to support research. Many companies offer competitive salaries and benefits, as well as career advancement opportunities. Scientists in the industry have access to cutting-edge technologies, state-of-the-art facilities, and plenty of resources to conduct their research. In addition, their work can have an immediate impact on society by driving innovation and contributing to the development of new products and services. Nevertheless, careers in the industry come with certain disadvantages. Profit and commercialization limit research scope, resulting in narrower opportunities than those found in academia. Scientists in the industry are subject to tight deadlines and commercial pressures, which may compromise the scientific integrity or shape the direction of their work. Furthermore, the hierarchical structure of industry organizations may limit autonomy and compel adherence to corporate objectives.

Government careers provide scientists with the opportunity to contribute to public policy, regulatory frameworks, and scientific decision-making. In govern-ment agencies such as national research institutes, environmental protection agencies, and health departments, scientists conduct research, provide expert advice, and shape policies related to scientific fields. A career in government provides scientists with a sense of public service as well as the opportunity to have a positive impact on society. Government positions often provide stability, job security, and competitive compensation packages with pensions. There are, however, some challenges associated with government careers in science. The implementation of many policies and programs can be hindered by bureaucratic processes, political constraints, and changing priorities.

UNITED STATES VERSUS INTERNATIONAL LABS

You can make better career decisions if you understand the similarities and differences between the United States and international lab settings. The funding landscape is one of the key differences between countries. Labs in the United States often rely on government grants, private donations, and corporate partnerships for funding. International labs, on the other hand, are subject to a variety of funding mechanisms, depending on the country's policies and priorities. In some countries, funding agencies support research well, whereas in others, funding agencies are quite limited.

In addition, labs in the United States generally have robust infrastructure and modern facilities. Through substantial investments, the country has developed a strong research ecosystem that allows access to cutting-edge equipment, advanced technologies, and specialized facilities. On the other hand, research infrastructure in other countries can vary considerably depending on their economic development and research investments.

Many United States laboratories have a strong culture of collaboration within and across institutions, which is crucial for scientific progress. A well-developed network of academic institutions facilitates connection between laboratories in the United States, fosters knowledge exchange, and accelerates scientific discovery. In other countries' labs, there are varying norms when it comes to collaboration and sometimes language barriers, although we believe that collaboration is an essential component to scientific research. Bringing diverse perspectives, sharing expertise, and pooling resources together through collaboration facilitates the solving of complex problems that transcend geographical boundaries.

CONCEPTUALIZING THE ACADEMIC PRINCIPAL INVESTIGATOR POSITION

A successful principal investigator (PI) in academia requires exceptional research skills, effective management capacity, and superb leadership abilities. Managing a laboratory and pursuing a career in science involve a wide range of responsibilities ranging from design of experiments to formulation of research hypotheses, development of methodologies, analysis of data, mentoring of students, allocation of resources, management of personnel, obtaining funding, and communicating and publicizing findings. Therefore, PIs must be highly organized, capable of multitasking and prioritizing their responsibilities, and strike a balance between their own research, administrative duties, managing resources effectively, mentoring, and personal affairs. By understanding key principles and implementing strategic approaches, individuals can enhance their chances of success in academic research.

Typical day-to-day tasks of an academic PI include keeping up with the latest advancements in science by reading the primary literature and meeting colleagues, identifying knowledge gaps that can be filled through novel research, and applying analytical skills to interpret complex data and determine meaningful conclusions. Meetings and discussions with the research team and collaborators are also highly important to monitor progress, troubleshoot challenges, and ensure that research is proceeding according to plan. In the background of all the aforementioned tasks, the PI must also keep track of funding and expenditures, which necessitates frequent grant applications.

Principal investigators need to develop a strategic plan for their lab and for advancing their careers in science. Publication of high-quality research articles in reputable journals is essential for advancement in academia. Funding opportunities must be closely monitored, grant proposals must be prepared often, and research must be funded in order to continue. Career advancement in academia just like other sectors is a longitudinal process that requires resilience, flexibility, and a willingness to embrace new challenges.

SETTING UP AN ACADEMIC LAB

As you establish your own lab, the major challenges will be purchasing and installing equipment, hiring and training your team, and obtaining the first few grants to establish your scientific niche. Many positions will require that you already have some grant funding prior to being hired for a faculty position. Although there are other tasks for newly appointed PIs, such as teaching and administrative duties, most institutions do not require new PIs to perform these duties during their first few years of service. Each lab requires different equipment and infrastructure, and it is imperative to be aware that in many cases maintaining the equipment is significantly more expensive than buying it (*e.g.*, look for a long-term contract that includes parts and technical assistance). Consider purchasing equipment with other PIs or core facilities within your department to share maintenance responsibilities. Make sure that funding is available to cover maintenance and training costs and ensure that the equipment is easy to use for lab personnel. You may wish to seek advice from colleagues who have similar equipment.

SUCCESSFUL ACADEMIC JOB OFFER NEGOTIATION

In the search for an academic faculty position, negotiating job offers is an essential but often overlooked step until it is suddenly time. Before starting a lab, consider physical laboratory space, a start-up funding package, access to essential equipment, teaching commitments, administrative responsibilities, and salary and benefits. Some of these aspects are not typically described in job postings and are highly dependent on the type of research program an applicant proposes. In general, negotiations are irregular, unstructured, and biased due to variability and lack of transparency.

During negotiations, keep in mind that you are seeking much needed resources for a laboratory to succeed and fulfill its goals. Break down the negotiation into several key questions so you will be better prepared for the process. In terms of scientific needs, what equipment and financial or other resources are required for your research program? How much and what type of teaching and academic service would you prefer to perform? Department chairs may encourage candidates to utilize shared resources rather than purchasing expensive equipment or release them from teaching duties to focus on grant applications during the start-up phase. What are your personal needs for success and fulfillment? What incentives will you need to attract talented individuals to join your lab? One of the most important assets of your laboratory will be its human capital. Do not forget to ensure that each employee receives the support and resources necessary to operate in the lab efficiently.

The purpose of negotiations for both the candidate and the employer is to agree on sufficient resources so that the candidate can be hired as a full-time employee and succeed in academic research. Job package negotiations typically involve identification of mismatches in expectations and proposing iterative solutions to meet and calibrate expectations. Candidates may also politely request that an institution match one or more aspects of competing offers if they have multiple offers, which is an excellent incentive to apply for and pursue multiple jobs in parallel.

SECURING FUNDING

Obtaining adequate start-up and long-term funding are the first and most challenging steps to establish a laboratory. Prepare a detailed budget that will assist in determining financial requirements, including expenses for equipment, supplies, salaries, and other operating costs. Consider sourcing funds from internal and external grants, investors, and fellowships. Set aside time to compose compelling grant applications to attract potential funding sources for the laboratory. This becomes a sort of business plan that should outline laboratory resources, objectives, impacts, and likelihood for success.

When submitting your application, many funding mechanisms and agencies require that you provide a substantial amount of preliminary data in order to convince the reviewers and agency that your grant hypothesis and objectives are feasible. However, it is quite challenging to obtain these data when setting up a lab (*e.g.*, buying equipment and hiring staff), establishing a teaching schedule (*e.g.*, building a new course is a highly challenging task; some people estimate that a lecturer needs about 20 hours of preparation for one hour of class), and setting up experimental systems (*e.g.*, specific animal models that may require many months of setup time). As you deal with these challenging tasks, there are two main approaches that you can utilize to obtain initial grants. The first is to obtain a substantial amount of preliminary data while you are still pursuing your postdoctoral training. Some principal investigators are less generous than others, but most recognize that you need to collect the data while you are doing postdoctoral research with their resources (*e.g.*, reagents, equipment, infrastructure, animal models, lab personnel, etc.). Alternatively, transition grants may provide an avenue to bridge this gap and generate preliminary data more independently, although these awards are requiring more and more preliminary data as well. There are early career transition grants in the United States (*e.g.*, National Institutes of Health (NIH) Research Career Development (K) Awards, National Science Foundation (NSF) Faculty Early Career Development Program (CAREER) awards, and other career development awards from private foundations and professional societies), but these grants are less common elsewhere.

State-of-the-art equipment drives high-impact scientific discoveries. Researchers around the world rely on cutting-edge instruments, devices, and equipment to make breakthroughs in their research. Thus, specialized equipment must often be acquired when starting a lab or entering a particular field, unless the lab space comes with equipment and/or the department maintains shared resources. Funding is required to purchase this equipment and also procure supplies (*e.g.*, chemicals, reagents, consumables, and biological materials) necessary for use of the equipment. New labs must also arrange infrastructure (lab benches, storage cabinets, etc.), utilities (waste, water, vacuum, gas, etc.), and safety measures (fume hoods, exhaust manifolds, personal protective equipment, etc.). Human capital is the next significant factor that must be funded and found to construct a successful lab. Budgets must support the salaries and benefits of the lab staff, including principal investigators, research associates, postdoctoral fellows, students, technicians, and administrative staff. A competitive compensation package is essential for attracting and retaining skilled employees who contribute to the success of the laboratory. Therefore, securing funding for the establishment of an academic laboratory requires new PIs to finance specialized equipment, supplies, furniture, utilities, and staff salaries. To procure and maintain these key ingredients requires the continual development of compelling research proposals, strategic negation of competitive job packages, and the ongoing use of multiple funding sources. For a more detailed discussion of grant applications, see Chapter 14.

Defining a Research Niche

Establishing an area of scientific focus as an early career PI helps with grant applications, career development, and reputation building in the scientific community. Good grant applications focus on a defined niche, which helps direct resources to generating adequate preliminary data on that topic and outline a realistic research plan appropriate in scope. Maintaining focus on a specific topic also allows early career investigators to make more progress in their field by consolidating their efforts into a main line of inquiry. Expenses can be consolidated since adjacent research projects within the same scientific niche can often leverage shared equipment, model systems, and supplies. Networking with colleagues more deeply in a specific field also helps improve the likelihood of favorable grant application reviews in that area. Once established in a particular field, it is much easier, albeit still challenging overall, to branch out into new domains and pursue new areas of inquiry.

ASSEMBLING A RESEARCH TEAM

Hiring and training the "right" individuals may be the most challenging task in establishing a new lab. There is no easy transition from doing things on your own and the way you want to do them to having other people do them in their own style (*e.g.*, storing reagents, analyzing data, presenting results, etc.). PIs are expected to spend most of their time writing grants, teaching, and handling administrative duties. Therefore, it is extremely important to thoughtfully assemble and train your team.

Before assembling a research team, establish your research objectives and determine the specific focus of your lab. Develop a set of core standard operating procedures (SOPs) for your lab protocols. Set a clear research agenda by clarifying your research interests, identifying knowledge gaps, and defining what you plan to do in terms of your work. This process will help attract talented researchers who share your passion and allow you to secure funding necessary to pay for those personnel and facilitate collaborations with other research groups.

Hiring individuals who possess enthusiasm, good work ethic, and the ability to learn and work with others is just as important as their existing tangible skills. Also consider alignment between employee interests and lab research priorities. Post job openings online via institutional and professional society websites, attend conferences, and reach out to colleagues and mentors for hiring recommendations.

Ask prospective lab members (whether trainee or employee) for professional recommendations, perform independent background research online, and conduct multiple interviews (*e.g.*, general professional interviews, interviews in front of existing lab members about the prior research or work that a candidate has performed, one-on-one interviews with a number of existing lab members, discussions regarding an article from the lab, etc.). When an applicant is selected, set clear expectations and give both parties (the individual and the lab) some adjustment time (usually 4–12 weeks, with regular check ins and informal constructive performance reviews), after which both parties should agree to continue working together.

Be thoughtful and intentional about how you will create a positive, collaborative work environment in the lab. Ensure that the workplace is a setting where employees feel valued and motivated by encouraging open communication, teamwork, and supportive working practices. There should be a clear understanding of what is expected in terms of work ethic, research integrity, professional conduct, and time management. As a PI, provide mentorship and professional development opportunities for employees, and promote work-life balance to achieve rigorous and realistic professional and personal goals. Regular group and individual meetings as well as informal electronic correspondence are essential communication methods to promote a high-functioning work environment. In addition, extracurricular lab social activities are an important mechanism to foster a cohesive and happy team culture.

ESTABLISHING LAB INFRASTRUCTURE

To carry out experiments and facilitate other aspects of research and to ensure the safety and productivity of team members, you will need a well-equipped and well-organized laboratory. Prior to setting up a laboratory, identify specific requirements of each research project. Consider the types of experiments to be conducted, the equipment needed, and the space necessary to accommodate the research team, which vary significantly between wet-lab experiments and computational work.

Choosing the right type of layout and design for your laboratory is imperative for productivity, safety, and collaboration in the workplace. Organize the lab space in accordance with the workflow of experiments. Separate areas for the preparation of samples, the operation of instruments, and the analysis of data. Keep RNA and DNA preparation benches apart from one another if these are processes to be utilized in your lab. Ensure that all institutional and governmental safety regulations and guidelines are followed for ventilation, waste management, personal protective equipment, etc.

The backbone of laboratory infrastructure is equipment and instruments. When choosing equipment and supplies, consider performance, maintenance requirements, reliability, and technical support. Assemble a network infrastructure that facilitates seamless data sharing and transfer between members of the lab as well as long-term record keeping for reference years in the future. Invest in hardware and software tools that will assist in the analysis, storage, and backup of data.

By determining research needs, designing lab layout, acquiring necessary equipment, developing robust data infrastructure, ensuring safety protocols, and collaborating with institutional support services, you can create an environment conducive to innovative research, collaboration, and scientific advancement.

LABORATORY MANAGEMENT

Successful laboratory management requires strong organizational and administrative skills. Setting realistic timelines and milestones according to the research plan and regularly evaluating progress are essential for maintaining productivity and promoting project success. Implementing efficient data management systems, adhering to ethical guidelines, and maintaining accurate records are also crucial for research integrity and reproducibility. Seek regular feedback and opportunities for improvement from lab members (no one should hesitate to speak up), and involve them in implementing these changes. Safety is of the utmost importance in any laboratory setting. Make sure you and your team are familiar with safety protocols and ensure that equipment and hazardous materials are properly trained. Prioritize regular safety inspections, emergency response plans, and the availability of necessary safety equipment to mitigate potential risks.

Collaboration and effective communication are essential for the success of academic labs just as any other workplace. Foster a collaborative atmosphere among lab members by encouraging open and transparent communication between them. Regular laboratory meetings, journal clubs, and scientific discussions can be used to facilitate knowledge sharing among the students as well as the development of critical thinking and group morale. Seek collaboration with other research groups to gain interdisciplinary insights and accelerate scientific research. Maintain relationships with other researchers and institutions within your field of expertise as well as those beyond your area of focus. These activities will increase chances for success and impact. Attending conferences, workshops, and seminars will give you the opportunity to share your work and learn from others. Collaborate with other laboratories on research projects, co-author papers, and apply for joint grants as a way of enhancing lab reach and visibility.

As a principal investigator, provide mentorship to your group members and facilitate professional development opportunities for their advancement. Encourage them to achieve success by providing them with regular feedback, guidance, as well as the opportunity to present their work at conferences and publish in reputable journals. To enhance their skills and expand their knowledge base, advocate for them to attend workshops, seminars, and other training programs. Ensure an inclusive and diverse work environment in which different perspectives are valued and equal opportunities are provided for all trainees and staff to succeed.

"PUBLISH OR PERISH"

The phrase "publish or perish" refers to the prevalent expectation that academic research scientists must consistently produce high-quality publications to maintain their positions and advance in their careers. Several factors have contributed to the development of this culture, including increased competition for limited funding, emphasis on journal metrics and rankings, and the desire for institutions to demonstrate their research productivity and excellence. Junior faculty members are often considered for tenure based largely upon their publication records (as well as other factors such as mentoring and teaching), and it is essential for tenure-track candidates to consistently produce high-quality research output. The pressure on researchers to produce publishable results is immense. The result is an accelerated work pace and a focus on quantity rather than quality. However, publication quantity over quality is inherently flawed, and there are also other valid measures of success and productively than journal publication.

Some aspects of "publish or perish" culture may drive productivity, but there are also challenges and drawbacks. The pressure to publish can create an environment where research misconduct or questionable practices occur, such as data manipulation or selective reporting, in order to enhance findings for publication. In addition, the emphasis on quantity over quality can lead to the production of incremental or less relevant research. Emphasis on publications can also overshadow other important aspects of academia, such as teaching, mentoring, and interdisciplinary collaboration. The "publish or perish" culture and its relationship with academic promotion significantly influence a scientist's career. By adopting strategic approaches to design research fit for both the advancement of science and for publication, while retaining a holistic view of their multifaceted roles in academia including mentoring and teaching in addition to research, academic researchers can navigate the challenges posed by the "publish or perish" culture.

SUMMARY

Starting an academic lab and research group is a challenging and highly rewarding endeavor. As a scientist, leading your own laboratory allows you to pursue your research interests, mentor students, and contribute to the advancement of knowledge in your field. However, being a principal investigator also comes with challenges and responsibilities that require careful planning and management. Carefully designing your research focus, obtaining funding early, and thoughtfully constructing your lab group will lay a strong foundation for success. Define your research objectives, apply for grants and awards, set up infrastructure, recruit talented team members, establish a supportive lab culture, provide mentorship and professional development, and build collaborations and networks. Remain adaptable and embrace change as new discoveries and technologies emerge. With dedication, perseverance, and effective leadership, your academic lab can make significant contributions to the scientific community and advance knowledge in your chosen field.

REFERENCES

Boyce, M., and R. J. Aguilera. "Preparing for Tenure at a Research-Intensive University." *BMC Proceedings, BioMed Central*, 2021.

Dolgin, E. "How to Start a Lab When Funds Are Tight." *Nature* 559, no. 7713 (2018): 291–294.

Goldstein, B., and P. Avasthi. "A Guide to Setting up and Managing a Lab at a Research-Intensive Institution." *BMC Proceedings* 15, no. 2 (2021): 8.

Greer, P. L., and M. A. Samuel. "Becoming a Principal Investigator: Designing and Navigating Your Academic Adventure." *Neuron* 103, no. 6 (2019): 959–963.

Kong, J. H., C. G. Vasquez, S. Agrawal, P. Malaney, M. M. Mikedis, A. B. Moffitt, L. von Diezmann, and C. M. Termini. "Creating Accessibility in Academic Negotiations." *Trends in Biochemical Sciences*, 48, no. 3 (2023): 203–210. 10.1016/j.tibs.2022.10.004

Liston, A., and S. Lesage. "Starting Your Independent Research Laboratory." *Stroke* 52, no. 8 (2021): e520–e522.

Masters, K. S., and P. K. Kreeger. "Ten Simple Rules for Developing a Mentor–Mentee Expectations Document." *PLoS Computational Biology* 13, no. 9 (2017): e1005709.

McKinley, K. L., A. L. Didychuk, D. A. Nicholas, and C. M. Termini. "The Transition Phase: Preparing to Launch a Laboratory." *Trends in Biochemical Sciences* 47, no. 10 (2022): 814–818. 10.1016/j.tibs.2022.05.002

Schwendinger-Schreck, J.. "You're Hired! Now What? A Guide for New Science Faculty." *The Yale Journal of Biology and Medicine* 85, no. 4 (2012): 563.

Somerville, L., et al. (2020). Three Keys to Launching Your Own Lab: Science| AAAS; 2019.

19 Alternate and Circuitous Paths

Bo'az Klartag

Perhaps it was hinted that I would become a mathematician already in 1964, quite a few years before I was actually born. On a certain night during that year, my grandfather chose to sleep on a park bench, even though he was a decent man with a respectable job (a tenured construction worker at Solel Boneh company) and with a comfortable bed in his home.

My grandfather was about to enroll his then 13-year-old son – my father – at a certain public high school. He slept on that park bench on that night because he was determined to be the first in line when the registration office opened in the morning. He wasn't entirely sure that if he took the earliest bus in the morning he would still arrive first, so he decided to hop on the last bus from Nesher to Haifa on the evening before. The plan worked perfectly, though he noted that quite a few slots were already reserved, maybe by parents who had better connections at the registration office.

This story is one of many that represent my family's attitude toward school and education. My four grandparents immigrated to Israel from Poland and from Tunisia, without much of a choice, accompanied by their children. They went through a complicated journey in their late 30s and early 40s and didn't look back to their countries of birth. They were off to a completely new start in a foreign country where jobs were scarce and whose language they barely spoke. My other grandfather, whom I am named after, passed away within a few years of arriving in Israel. It was my uncle who, as a teenager, worked during the day and studied at night in order to support the family.

All evidence suggests that there was a strong sense of commitment and sacrifice on behalf of my grandparents (and uncle) in order to allow my parents to concentrate on their high school studies. Thus, they had to excel, if only to justify the nearly heroic efforts that surrounded their going to school. My father was a "wunderkind." He skipped a grade in elementary school – the first child to ever do so in the history of his town – and went on to study electrical engineering at the Technion. My mother studied mathematics, also at the Technion.

One evening, toward the end of my mother's first year at the Technion, the principal of the high school from which she graduated knocked on the door of their apartment. He offered her a job, teaching mathematics at Kiryat Haim High School, merely a year after her high school graduation from the same school. She agreed, and before her 19th birthday, she became a high school math teacher, parallel to her

DOI: 10.1201/9781003301400-19

academic studies. She taught math in high schools for 45 consecutive years with devotion, not taking any break or sabbatical leave, even at tough times.

I am the elder son in my family. In 1985, when I was seven years old, I received my first personal computer. Before getting an actual computer, I had played and pushed random buttons on two oscilloscopes that I got while in kindergarten(!). My father somehow brought these home from his work. Yet they were nothing compared to my 128Kb Amstrad computer. While I don't know what my first words in life were, which were surely in Hebrew, I definitely remember the first words that I learned in English: run "menu." This is the BASIC command needed in order to run a certain menu of games. The Amstrad brand wasn't a great success, and there weren't so many computer games available to me. Thus, I had to learn to program in BASIC, using a steady supply of booklets that I received from my father.

When I was in third grade, a few kids who liked arithmetic were invited to attend a certain afternoon class, taught by an energetic teacher named Gabi Yekuel, at my elementary school in Kiryat Bialik. I still have my notebooks from third grade. He taught us how to solve two linear equations in two unknowns. Speed was considered of great value, and the first child to announce the correct value of X and Y would win a chocolate log ("Mekupelet"). To this day, I feel conditioned to crave this type of treat whenever I solve a math riddle, though the level of the riddles that I am asked to solve has varied since those times.

By the time we began learning the English alphabet at school in fourth grade, I already had a few years' experience of typing in English on the computer – all of these *goto* commands, *for* loops, you name it. I simply knew that the future lies in typing and not in handwriting. To this day I regret my attitude back then; my handwriting is awful, while as a mathematician I am required to write quite a lot on paper with pen and on blackboard with chalk or marker. There was also the question of the correct Latin spelling of my first name. While the canonical spelling "Boaz" appears in the King James version of the Bible, among other respectable sources, my cousin's computer had a software that would read out loud whatever you typed. It was decided that Bo'az sounds closer to the Hebrew pronunciation than Boaz, so we went with that.

While I liked my computer and participated in a weekly enrichment math class after school, the most important thing for me as a child was soccer. I played soccer with my cousin's friends virtually every day, and the success of the Maccabi Haifa football club was of paramount importance to me. I also read a lot as a young child. I think that I read most of the books in my school's library, including a 17-volume encyclopedia Aviv. I have worn glasses since age 8, and my friends told me this meant that I would never become a pilot in the air force, as desired by many Israeli boys at the time. In short, I had a completely normal and happy childhood.

A transformative event in the development of my worldview occurred during the summer vacation between sixth and seventh grade. I was too old for a little kids' summer camp, and my parents sent me to a ten-day workshop of scientific experiments at the youth division of Tel-Aviv University. I was even allowed to take the bus back home by myself and felt rather mature and independent.

The teacher running the workshop was named Nathan Arie. He arrived at the first lesson holding a basketball, and we began passing it around to one another. After a few minutes, we all agreed that there is this strange thing called *Inertia*, as otherwise nothing would make the ball continue to move in the air. We had experiments demonstrating Newton's laws, and we even tried to conduct Millikan's famous oil-drop experiment, trying to measure the charge of a single electron (some kids claimed to see something through the microscope, but I saw nothing).

This workshop was a revelation for me. Newton's system of the world was the decisive, clear explanation of what is going on. Anything else paled in comparison. I just knew that when I grew up, I will become a physicist (spoiler: I did not, but I am not that far from it). Everything had a meaning. The important thing in life was to unveil the secrets of the world through science. It was during this summer that I officially became a nerd. While in sixth grade I played with friends and went to school parties. In seventh grade, I watched in horror as my classmates began to smoke cigarettes at the junior high school.

I am sure that it wasn't only because of these ten days in the summer holiday, but somehow my attitude to the general society became somewhat that of an outsider: they were "muggles" who knew nothing about real stuff like science. I surely still played basketball for Maccabi Ra'annana throughout seventh grade, but I felt that my interests in life had evolved. At home, we had some popular science books, by Feynman and by Weisskopf, and I tried my best to understand them. I registered for the winter classes at the youth division of Tel-Aviv University, where we were supposed to hear about something deep and mysterious called quantum mechanics, among other topics. These winter classes were in fact a disappointment to me. I didn't understand many things that the teacher said. Whenever I would raise my hand and say something about physics in class, it wasn't accurate. Still, at the end of every lesson, the teacher would give us a math problem. These math problems were quite to my liking, actually.

Perhaps I liked Newtonian mechanics just because it was so mathematically elegant, and because it saliently implied that "math runs the world." In retrospect, it seems ridiculous to try to understand quantum mechanics in seventh grade, without knowing about complex numbers for example. The next year, while in eighth grade, I took the mathematics winter classes for youth at Tel-Aviv University. In a sense, this class for curious, math-oriented kids essentially put me on the track where I find myself today. The teacher in this class was Haim Shapira.

At some point during the course, Haim asked us the following question: how many ways are there to climb a ladder with 16 steps, if you can climb either one step or two steps at a time? He said that whoever solves this question should go on to study math as a real student at the university. I still remember my answer, 1597, which I reached via complicated computations, involving dividing the total count into quite a few cases. I was surprised to see that my final answer was correct, with so many potential pitfalls around. When I presented my solution on the board, Haim would stop me and tell me things like: you know, the formula you wrote down is called "n choose k." Then two other kids described another solution: they replaced 16 by smaller numbers, inspected the beginning of the sequence, and noted that this is a Fibonacci sequence. I felt like an idiot. I worked so hard and didn't notice this easy solution! Still, the teacher seemed to like my ignorant approach. He said that I should just remember this little trick, playing with small numbers first.

Haim suggested that I join a certain experimental program, where kids of my age take a preparatory class once a week in the afternoon and then enroll in standard university math classes. He was the teacher – as well as my personal guru at the time – and I liked this class a lot. He had a subversive attitude toward the way math is taught in "regular" schools. He begged us not to listen to the combinatorics lessons in the usual school, claiming that this will destroy our mathematical intuition. He taught us trigonometry in 20 minutes, just the definition of sine and cosine, with the homework assignment for the vacation being: prove all formulae in the list on the last page of the high school book. Not everyone passed the class; it was only me and an older kid. Thus, it happened that by ninth grade I was allowed to enroll in regular math classes with adult students at Tel-Aviv University.

Today things are different. We live in the age of information, and many kids are exposed to rather advanced math and science (or at least the executive summary thereof) just by watching videos on the Internet. In the early 1990s, however, there were almost no school kids who knew advanced math or took regular math classes at the university. Most Israeli math students had already completed their military service of three years, so they were mostly in their 20s. I remember that the female students were particularly kind to me, as a child that was sent into the world of adults. One of them showed me the university library, and I felt like I was entering the temple of knowledge, with so many new books for me to read.

These were also the days of the Russian immigration to Israel, and it didn't take too much time before these talented, hard-working immigrants took over the young math scene in Israel. In 1996, I flew to India as a member of the Israeli team for the International Mathematics Olympiad (IMO). All five of the other team members were born in the Soviet Union and were of relatively modest economic background in Israel. I was ranked third in the team. It was not easy to compete with kids who attended mathematically oriented schools such as Moscow school 57 or St. Petersburg school 239.

I felt a deep appreciation for these highly educated Jewish immigrants from Russia. Some of the best teachers of my undergraduate studies were Milman, Olevskii, and Polterovich, who were such immigrants. In my first calculus lesson at the university, I remember that Rosenthal told us that we were going to learn exotic things about limits and infinities and real numbers made of Dedekind cuts, but that we should always remember one thing: we never divide by zero.

I graduated from Tel-Aviv University in 1997, with a major in mathematics and a minor in computer science, and I still regret not taking physics classes as well. I was drafted to the army upon graduating. Two years later, when I was still a soldier, a friend of mine told me that Milman asked him to invite me to come to his seminar. I was very flattered that Milman remembered me, and promptly showed up to the seminar. Afterward, I spoke to Milman and told him how much I enjoyed his Functional Analysis class. Milman answered immediately, that there are too many counter examples in this field, and nowadays one should study finite-dimensional spaces instead. I had no idea what he was talking about. Still, he handed me a paper by Bourgain, Lindenstrauss, and himself, and he asked me to read it, with an invitation to return and discuss the paper with him.

It was a difficult paper for me. I distinctly remember how I struggled and completely rewrote it "in my own words." Milman asked me if one can somehow modify the proof from the paper and obtain a formula for the number of random symmetrization steps needed to transform a convex body into an approximate Euclidean ball. I must say that it was quite easy. The answer was almost written in the paper itself. Milman was pleased with my answer, and he asked me to talk about it in his seminar and said that this is almost a master's thesis. This was only a few weeks into our regular meeting sessions. I thus became his MSc student, and he asked me another math question on symmetrizations. This one was not so easy. In fact, I was stuck for many months and became desperate.

In retrospect, I learned a lot of mathematics while trying various directions of tackling this specific problem. It was an effective way to study math – different from the undergraduate way, where you learn a subject just because you were told to do so. Here, I learned quite a few math topics with the hope of either using them or quickly ruling out their potential use. Therefore, I automatically focused on the heart of the matter, leaving technical details for later. Had I been able to solve this question earlier, I'm sure that I would have ended up knowing less math.

Some months went by. I finally gave up and asked Milman for his approval to submit the results I had then as a thesis. I also requested that we meet only once every two weeks, and not weekly as before. The math problem was too hard for me. I was stuck for too much time and felt frustrated. Milman said that my words give him "pain in the heart," but that he could not force me to stay. I submitted my MSc thesis, published it as a mediocre paper at Milman's seminar proceedings, and stopped meeting with him. Still, at the MSc graduation ceremony, I gave a speech as the representative of the graduating class. I decided to try physics again, and so I read a math/physics book by Arnold, which I thought was a good place to begin learning physics, at least for a person with a mathematical background like myself.

At this point, perhaps ten months after giving up on that math problem, I somehow got an idea about how to solve it. I don't know exactly how it happened – it wasn't a radically different idea from what I had tried earlier. One could say that it really helped to take a break from the problem and then return to it later on. But I actually blamed my girlfriend for this eureka moment. It seemed that all of my problems were miraculously solved when I met her in April 2001, including that formidable math question that had me stuck for so long. We're still together. I remember writing down the details of the solution on my computer until 3 AM, and then again until the same hour the next day, and after a few more days like this, I thought that I finally had a complete, presentable solution.

I came back to Milman, who was very happy about this development. He asked me to travel to the island of Crete in a few weeks time, with all expenses paid by his research grant, and give a lecture about my result at some workshop that he organized there. In Crete, I was introduced to several mathematicians, whom I knew only through their papers, and some of them expressed clear interest in my result. It seemed obvious that my next step would be starting my PhD studies with Milman.

In my PhD studies, I continued to study symmetrization processes. I also became obsessed with a mathematical question called the *slicing problem*, after my adviser introduced me to the brilliant mathematician Jean Bourgain. As of May 2022, this problem is still not completely resolved, though we now know much more than we knew 20 years ago. In the third year of my PhD studies, I felt that it might be time for the next step. I applied for a post-doctoral position at only one place – the Institute for Advanced Study in Princeton where Bourgain worked. I was very fortunate to have a particularly caring PhD adviser, who uncommonly arranged for recommendation letters for me without telling me a thing, in order not to divert my attention from math itself.

I was accepted, and readily moved to Princeton, which for me was the town of mathematical living legends. Well, I got depressed rather quickly. The mathematical life there was not organized around a seminar as it was in Tel-Aviv, at least for me. This gave me a sense of emptiness, so I tried to organize a seminar with other postdocs, but it waned away after a few meetings. I also realized that Bourgain's interests had by then drifted away from the slicing problem. I tried to work on new directions, to break my scientific isolation. I took some classes at Princeton University – I always loved taking math classes – and became hooked on a class about the Whitney extension problem by Charlie Fefferman (speaking about living legends, at some point he was the youngest professor in US history).

My class with Fefferman felt like a "master class." I felt privileged to learn and watch closely how a great mathematician works. Only two students, myself included, made it to the second semester. At some point, Fefferman mentioned a certain open problem, which was quite appealing to me. The other classmate, Nahum Zobin, and I started thinking about it together. It went well, and toward the end of the semester, I lectured about our solution. This was my first taste of success since arriving at Princeton.

I consider the following year, still as a postdoc at Princeton, the most successful year of my entire career so far (though I am not sure if in general a mathematician's trajectory should be called a career). During that year I worked both with Fefferman and alone, studied, solved, published, and understood. It was almost too good to be true. I felt that I could successfully contribute to more than one sub-field of mathematics. A few years later I even received some prizes for my work and became an established mathematician.

After four memorable years in Princeton (the last two as an assistant professor), in 2008 I returned to my alma mater, Tel-Aviv University, as a faculty member. My parents and siblings were all waiting for us at the airport upon our return home to Israel. My sister was holding a helium balloon. My grandfather was surely happy and proud too, though he didn't come to greet us at the airport, being 96 years old. We saw him a few days later though, and he smiled as he played with our little baby.

20 Navigating Decisions, Milestones, and Crossroads in Science

Samuel J. S. Rubin and Nir Qvit

AVOIDING COMMON PITFALLS AND INEFFICIENCIES

Much of our journeys have consisted of trial and error. In that process, we have made countless missteps and mistakes, but we have also avoided mistakes by observing others and learning from their ways. Here, we translate these experiences into key lessons, which we hope will allow future trainees to avoid common pitfalls and inefficiencies in their scientific journeys. In short, we made the mistakes or observed a colleague making the mistake, so you don't have to. Of course, there is nothing wrong with learning from error and we highly encourage this practice, but if we can save future scientists a bit of time, effort, and hardship then we will be glad to have shared the following.

Assert Yourself

One of the most frequent lapses in judgment of the novice is failing to assert oneself for fears of overstepping academic hierarchy, fears of asking too much, or assuming that an individual would have no interest or willingness in addressing the individual. The most common manifestation of this phenomenon is failing to ask for help. Even if an individual is not normally shy or lacking in confidence, the power differential in academic and professional settings can often lead one to back down from asserting themselves. Most personalities in academia want to help trainees – regardless of how rough around the edges they may seem. Introduce yourself to people no matter how powerful they may be. Make cold calls. Send cold emails. Ask for help or guidance. Politely present your proposal or request. At the least, you will have a new connection and learn something in the process. In the best-case scenario, you may obtain exactly what you sought or more.

A colleague Sarah was composing her graduate school application essays. She agonized over how to present herself. There were too many permutations and no single way to tell her story. She spent weeks writing and re-writing, consuming countless hours of her time. She could not imagine bothering any of her professors to review her now numerous drafts and help her frame a coherent argument for her application. Finally, a classmate suggested that she bite the bullet and ask Professor

DOI: 10.1201/9781003301400-20

Altoviti if he would be willing to meet to help provide some guidance on her application essays. To her surprise, Professor Altoviti responded to Sarah's email within minutes, explaining that he had been on multiple graduate school admissions committees in the past and would be happy to meet with her after she sends him a draft to review. Their meeting led to a longitudinal mentorship, during which Professor Altoviti became a strong advocate on Sarah's behalf. She didn't expect this outcome, and neither should you – but now you know that it's possible if you take initiative and ask.

LEVERAGE YOUR NETWORK

There are three common reasons that people neglect leveraging their network of individuals for support when pursuing a goal. First, doing so effectively requires energy and effort with the hope but no guarantee of a positive outcome. This is flawed because when practiced in good faith with kindness and compassion, networking more often than not produces gains sooner or later. Second, some individuals feel that they must earn an achievement independently without the support of others to reach the goal more efficiently or successfully. This is an arrogant misconception that anything earned with the help of others is not a genuine achievement, when in fact the best accomplishments are made collaboratively. Reaching a goal with support from your network in no way diminishes your achievement, but rather enhances your success in sharing the experience with those who helped you realize it for yourself. Third, other individuals feel that it is not fair for them to call upon their network to assist in reaching a goal when others are not as fortunate. Hiding behind this humble facade is simply unrealistic, as even the most unconnected individuals have or can identify individuals to call on for support. In short, most people leverage their networks when striving for goals. In doing so, you will accomplish more faster, and it does not detract from the fact that you earn your achievements. If you fail to leverage your network, someone else will, and at some point in the future, you will regret not taking advantage of this opportunity.

GO STRAIGHT TO THE TOP

When asking for permission, endorsement, or assistance, the naive individual tends to approach another individual with status closest to their own in the academic or professional hierarchy. This is a mistake because policies and decisions are usually set from the top down. Contrary to what one might expect, this slightly superior individual who is subordinate to someone else is more likely to make a conservative decision in order to please their superiors, which is less likely to be advantageous for you. In this case, it is poor form to then approach a superior individual to pose the same request, as this can be interpreted as disrespectful to the authority of the former individual and the leadership of the latter. Thus, it is most advantageous to prepare a strong rationale, gather whatever support you can, and then approach the top decision-maker with authority to grant your request.

Recently, I failed to follow this advice, and it prevented me from meeting my objective. After having completed extensive remote virtual interviews and meetings, I was considering moving to another institution. However, I was still undecided about the move. I felt that a site visit in person to the facilities and individuals on the ground would help me determine if the new institution would be the right fit for me. In my mistaken logic, I asked the individual in a leadership position who I had most recently interacted with in virtual interviews if I could arrange a site visit. She politely explained that since all interviews were conducted remotely, no candidates were allowed to visit in person, since a site visit could be interpreted as a demonstration of interest and that this disadvantages candidates without the resources to easily make multiple site visits. What I neglected to realize is that there was an individual in a higher leadership position more invested in my interest and with greater decision-making authority who would have gladly coordinated a site visit on my behalf. However, once I had already been declined by the individual I approached, it would have been inappropriate to supersede her authority by making the same request of her superior.

SET EXPECTATIONS AND MAINTAIN COMMUNICATION

Ignorant or perhaps hopeful individuals sometimes fail to clearly set expectations up front and maintain consistent communication when entering a project, working relationship, or other interpersonal effort. This can be convenient when the full context and extent of work have yet to be defined and when greater flexibility is desired. However, time and time again we have seen later conflict arise due to differences in the simplest assumptions made by parties earlier on. To the extent possible, it is therefore imperative to agree upon responsibilities and expectations – or lack thereof – at the earliest signs of individual investment. There is usually ample opportunity to adjust and realign expectations as the enterprise evolves, and for that reason, it is just as important to maintain consistent and open communications about individual responsibilities and progress. Regular mutual feedback will reliably help avoid miscommunications and misalignment of expected outcomes. Putting in the effort to maintain regular communication will save potential headache later.

RECOGNIZE AND ADDRESS RED FLAGS

Ignoring red flag warning signs in projects or individuals is dangerous. Hoping these issues will resolve spontaneously is delusional. While it can take time and experience to learn to identify red flags on your own, it is important to act when they arise. Either address issues directly and to your realistic standards or start planning an exit strategy. Once an interpersonal or technical problem arises, it is unlikely to resolve spontaneously without intervention. These problems generally foreshadow more problems to come unless efforts are directed toward identification and mitigation of the underlying issue(s). Alternatively, if losses are to be cut, it is better to do so earlier rather than later when more liabilities are likely to be involved.

At one point in his training, Sam had discovered a novel therapeutic compound and platform technology with the potential to revolutionize a significant portion of the biologic industry. With his advisor, Sam filed a patent application, and they decided to create a startup company to commercialize the technology. In the process of accelerating research on the therapeutic lead and preparing the technology for commercialization, Sam proposed a method to modify the compound such that it could cure the disease indication in addition to treating the symptoms, meaning that if successful then after a certain period of time the therapy could be withdrawn and the patient would retain lasting benefit. Sam's advisor, with significant prior experience in the biotech startup industry, was quick to point out that this proposed cure was not a desirable business approach because it would be more beneficial for commercialization purposes if the lead compound remains a therapeutic but not curative treatment so that patients would need to take a medication for the rest of their lives and thus generate revenue for a longer period of time. Sam was appalled at the thought of sacrificing his motivation to improve patient lives for financial interest, but he was ultimately convinced by his advisor that this approach was required in order to establish commercial success before expanding to other permutations such as curative therapy. He figured, hopefully, that working toward a technological and financial success in this first iteration would later allow him to pursue novelties such as the curative iteration. It turned out that this bump in the road was a harbinger for much worse to come. In short, the entire start-up effort ended in disaster due to breach of trust.

MAKE ADJUSTMENTS

Once a project or process has begun, it can quickly feel too late to make adjustments. This section is about how it is never too late to make adjustments – minor or major – when they have the potential to make significant impacts on the quality of your experience or the outcome of your efforts. As with many of the lessons discussed here, it is better to act sooner than later. When well-coordinated and impactful, it is always worth putting in the effort up front for the long-term relief. If something causes burden or frustration in the present, it is unlikely to be tolerable or sustainable in the future. The long-term satisfaction is worth a little discomfort to make changes while there is still time. This could be in the form of adjusting personnel involved in a project, redirecting a project objective, switching programs, or other scenarios. In any of these circumstances, toughing it out and powering through is rarely worthwhile unless you are already near the destination and success is highly likely (*e.g.*, one term away from finishing a degree).

CONFIRM KEY AGREEMENTS IN WRITING

It is unnecessary to form all agreements in writing, but it is incredibly helpful to confirm key agreements in writing. This does not need to occur via formal channels in every case. Often, a simple follow-up email summarizing a discussion or meeting is appropriate. This ensures all parties are on the same page, gives opportunity for clarification or realignment, and allows everyone involved to be held accountable equally and objectively. Circumstances where this is especially critical include authorship on publications, responsibility for key tasks, and expected timelines for project milestones. Other cases where formal legal documentation of an agreement is warranted include patent applications, confidentiality agreements, and commercial and financial documents. Never sign a legal document without reading and understanding it in its entirety. If this is not possible, then do not sign the document and/or seek legal counsel. Large academic institutions often provide free legal counsel for trainees, and many attorneys or law firms also offer limited pro bono services as part of their required duties.

Under Promise and Over Deliver

A general best practice is to under promise and over deliver whenever possible. This approach is most likely to result in positive feedback on your performance while maximizing your agency. Since unanticipated challenges and opportunities arise, it is advantageous to retain some flexibility in how you dedicate your time and effort. Agreeing to assume responsibility for an undertaking before fully understanding the requirements involved and the realistic chance of success can also be problematic if expectations are not met. It is much better to exceed expectations than to fall short.

An example of politely under promising and over delivering can be found in a professor asking a student to compose a manuscript to submit for an invited review article. Often more senior academicians will be invited to contribute review articles on a particular topic. Since they may not have the bandwidth to compose the article themselves, they often ask a trainee to compose a draft. This can be advantageous for the trainee in some cases when they stand to benefit from creating a publication and from reviewing the literature on topic for their own interest. However, such a project can also be a distraction from other more academically lucrative efforts. A polite way to address this scenario as the trainee is to thank the professor for thinking of you, express some interest in the opportunity, explain that you have several other obligations, suggest that you may be able to complete a portion of the tasks in a certain timeframe if your time permits, and ask for additional information on expected contributions and timeline. This demonstrates your appreciation and your consideration in taking the proposal seriously, but it does not commit you to fulfilling specific duties unless you subsequently explicitly agree to them.

Preserve Your Network

The scientific community is smaller than one might imagine, and it is important to preserve your professional network without burning any bridges because it is impossible to predict when paths might cross in the future. Specifically, it can be particularly helpful and also gratifying to keep in touch with former mentors, which is often overlooked by trainees as they advance through their careers. This practice credits the impact of your teachers and the meaning of their work and also engages these individuals in your success. Mentors end up in such positions because they are invested in their mentees' success, and thus these individuals are important to retain in your closest network of supporters. It is unnecessary to maintain frequent contact with each of your many teachers, but do reach out periodically to as many as possible.

Hold Your Cards Close to Your Chest

While it is a valuable part of the research process to discuss findings and new ideas with colleagues, there exist just as in any field competitors and individuals with malintent. At conferences and venues with larger or unknown audiences, be circumspect about how much unpublished or non-public information you share and with whom. There is still plenty of discussion to be had without revealing all the details of your latest work. Several colleagues have been "scooped" by another researcher who learned of their methods and findings, used this information to replicate experiments, and then published the work first. It is unfortunate that this can happen in what would ideally be a collaborative field, but the reality is that getting scooped can be a concern in competitive areas. Of course, when surrounded by trustworthy colleagues at a lab meeting or departmental seminar, the benefits of obtaining helpful feedback, questions, and new perspectives usually outweigh the risks of full disclosure.

Prioritize Balance in Your Life

Scientists are notorious for intense work ethics and maintaining demanding work hours. For many, this is sustainable for short-term courses surrounding deadlines or exciting periods of progress. But for many, this is not sustainable in the long term. It is too easy to get lost in the never-ending saga of a good scientific research question to the point that other aspects of life become compromised. If your life balance is not adequately prioritized, then eventually your work and your well-being will suffer. We favor a preventative rather than reactive approach to this matter. Schedule vacations appropriately, maintain social connections, maintain physical activity, prioritize good sleep hygiene and nutrition, and execute any other practices that contribute to your well-being. Ultimately, your science will benefit too.

A SYSTEMATIC APPROACH

We recommend a systematic approach to most consequential choices and decision points. The first step is to thoroughly define the problem, question, or decision to be made. This includes seeking out as much information as possible about the context. We hope and have generally found that most people in science are well disposed toward helping fellow scientists and trainees. However, the reality is that people also often stand to gain something when offering an opportunity. Thus, it is important to seek additional information and perspectives from others with more or different experience in order to understand the full background of the situation.

Next, define your desired goal to be achieved after you traverse the crossroad or milestone. This objective should be individualized and matched to your own priorities. Then, begin identifying potential solutions to address your specific goal. Accomplish this on your own and in consultation with trusted confidants. Think outside of the box. Pursue other perspectives. Don't limit your options yet, even if something doesn't strike you as the most realistic. Once you have identified multiple potential solutions, it is time to carefully assess the effects and possible implications of each. Consider immediate and long-term ramifications. Conduct a mental exercise where you fully immerse yourself in each potential solution and envision the possible outcomes. Pay close attention to whether the rationale is logical, the prospect of success, and your gut feeling about each option.

Once you have gathered this information, prepare yourself to choose a path forward. Sometimes there are wrong choices, but usually there is more than one correct choice. Pick one and unless making that choice sends you clear signals that it is not the best fit for you, then commit to that choice and proceed to execute the plan. It is not too late to switch directions when appropriate conditions are met. In order to make progress in anything, you must trust your gut and your research in making a decision and committing to that plan.

SWITCHING PROJECTS

A research project should be chosen with great care and meticulous planning. Nevertheless, there are certain instances where realigning your effort in a different direction or with an entirely different project is appropriate. Scientific research is tremendously challenging, as it often involves a process of educated trial and error with frequent, repeated failure. This cycle can easily become exhausting and frustrating to the most talented of investigators, causing some to live on the brink of burnout. It is important to clearly identify and be in touch with your motivation in order to persevere and navigate these longitudinal challenges to progress. We review in other chapters how to set up experiments such that no matter the outcome, whether failure or success, you learn something useful for your next step. When troubleshooting failures and approaching enigmatic problems, exhaust your network and other resources to help overcome obstacles. Fight for your ideas and your projects. There aren't many easy discoveries.

When you, your advisors, confidants, and colleagues form some sort of consensus that continued pursuit of a particular project or path is extremely unlikely to succeed to the point of bordering on absurdity and pointlessness, then it may be time to realign your efforts in order to conserve time and resources. In this unusual event, ensure that all stakeholders are on the same page, and engage each in salvaging as much as possible from the project before abandonment. Simultaneously begin planning your next steps as soon as such a pivot becomes evident. Determine whether a publication or other commodity can be obtained from whatever was accomplished in the project and tie up any loose ends to the extent possible. Failure is normal in science, but it is important not to abandon ship too early and mistake challenge for failure.

SWITCHING ADVISORS

Partnering with an advisor is a decision that should be made with careful consideration and mutually agreed expectations. There are rare cases when switching or disengaging from an advisor may be the best path forward. When even remotely considering this option, one must proceed with great caution due to the potential for bruised egos and political ramifications. If you are wondering whether to continue a student-advisor relationship, then you are in need of additional mentorship. The first priority should be identifying independent, trusted mentors to help support and guide you through this process. Also ensure that you have made every best faith effort to reconcile whatever differences arose between you and the initial advisor. Make sure to document your efforts so there is no question about whether or not you made a good faith effort.

When broaching the topic of disengagement with the individual from whom you intend to disengage – whether via de-escalation or ending of the relationship – first arrange an appropriate time and place to begin the discussion. Try to choose a neutral location and an area with some privacy but not so much seclusion that it would be difficult to extricate yourself if the interaction becomes confrontational. Give the other individual an opportunity to initiate the disengagement first; chances are they feel similarly, and it will go more smoothly if they think the disengagement is their idea. If they don't take the hint, then express your gratitude for all they have taught you and focus your discussion of the topic on why an alternative arrangement is a better fit for you and ultimately them too. Rather than focus on your differences or their deficiencies, explain why a different arrangement is in both your best interest, without going into too much detail about your subsequent plans. Keep the interaction as short and straightforward as possible, and make sure that by the end of the discussion you have reached a mutual understanding. As usual, follow this up with a short email summarizing the plan on the record.

PIVOTING TO A NEW FIELD OR SETTING

Redirecting your focus to a new topic or even transitioning to a completely different work environment can be a daunting thought. These shifts are often necessitated by factors difficult to control and can be an exciting, nerve-racking opportunity to expand your horizons. Challenge is certain, but making these transitions is worth the effort if you are motivated and the destination inspires your creativity. A common concern for individuals making a transition to a seemingly foreign domain is that they will be somehow underprepared or inadequate to function at a high level in the new setting. In contrast, it is often the unique perspectives that an individual brings to a new setting which allows them to excel. Diverse expertise is a marketable feature and demonstrably advantageous for problem solving, even if not traditionally applied to a given domain. For instance, artists and sculptors bring pragmatic skills to the engineering arena that can prove instrumental for solving the most challenging problems. Many of the most transformative discoveries and applications were made by individuals who followed their curiosity into new areas of practice.

A BASIC INTRODUCTION TO LAW FOR SCIENTISTS

There are several common situations in the academic investigator's career when a basic understanding of legal mechanisms is helpful. Neither of us are trained in law nor purport to be lawyers. However, we hope that sharing our experience with the following common cases will help you navigate some of these more frequent occurrences. We suggest that whenever possible it is still best to consult a legal professional – but we realize this is not always realistic. Some of the fundamental principles that help avoid legal problems have been mentioned elsewhere in this chapter and this book, but generally include documenting in writing any agreement of consequence (regardless of how much you trust or how well you know the individuals involved) and fully reading and understanding any document before signing it. In addition, a great appreciation for the law and the influence of the legal system is required to avoid running into problems. It is important to realize that there is much to do legally that requires years of experience to fully understand, and thus it can be tremendously helpful to consult an expert if and when a legal matter arises.

MATERIAL TRANSFER AGREEMENTS

Material transfer agreements (MTAs) are legal documents that concern the exchange of materials between two or more parties. Common materials include data and research reagents, such as chemicals, genetic material, proteins, cell lines, animals, etc. An MTA typically defines the rights and obligations of each party, such as how the materials may be used, who can publish what and when, and who owns derivative intellectual property. MTAs can range from a simple letter to a complex document depending on the context and whether intellectual property is involved. Many institutions offer standardized MTAs, although for some cases it is necessary to customize the document. In an academic setting, the most frequent application of an MTA is to the exchange of materials between collaborators at different institutions, in which official institutional representatives may be required to sign the document.

CONFIDENTIALITY AGREEMENTS

Confidentiality agreements are frequently referred to as non-disclosure agreements (NDAs) or confidential disclosure agreements (CDAs). NDA is the most commonly used term, sometimes preferred in the setting of transactional matters. CDA is also used quite often, sometimes preferred for non-transactional matters, such as consulting or the performance of services. NDA and CDA are essentially the same in principle; the content of a confidentiality agreement is what sets it apart rather than the name.

Generally, a confidentiality agreement represents a contract between two or more parties that governs the exchange of information or material for specific purposes with certain conditions or restrictions set for the use or dissemination of that information. This type of document can be used to protect trade secrets, intellectual property, and other proprietary information. Confidentiality agreements can be mutual, or restrictions may only apply to one party involved. A unilateral or one-way agreement concerns the exchange of information from one party to another. A bilateral or two-way agreement concerns the exchange of information between two parties each way. Multilateral agreements can be used when more than two parties are involved. Confidentiality agreements are commonly invoked when information must be shared with new parties in the process of using a scientific discovery for potential commercial applications.

INTELLECTUAL PROPERTY AND PATENTS

Intellectual property (IP) refers to human creations or inventions. In the United States, IP includes copyrights, trademarks, trade secrets, and patents, although the legal definition of IP various by country or legal system. Patents are by far the most relevant and commonly used form of IP in science (Murphy et al., 2015). A patent allows the owner authority to try to exclude others from making, using, or selling an invention for a limited period of time (usually 20 years from the filing date) in exchange for publishing details of the invention. A patent also prevents someone else from obtaining a patent for the same invention. In the past, the United States Patent and Trademark Office (USPTO) could grant a patent to individuals who demonstrated they were the first to invent; the America Invents Act of 2013 changed this system to align with most foreign patent systems such that patents are now granted to inventors who were first to file a patent application, with a few exceptions beyond the scope of this discussion. Infringement on rights afforded by a patent occurs when another party makes, uses, or sells the invention without authorization from the owner. Infringement is enforced on a national level, so the same rights may not be afforded in another country unless a patent is filed in that jurisdiction. Since patents usually fall under private law, a patent owner generally sues another entity in civil court for infringement in order to enforce their rights.

Patent applications are submitted to and granted by the USPTO. Only inventor(s) can apply for a patent, but ownership of granted patents or certain rights can be assigned to another entity. Inventors who made their discovery in the course of work at an academic institution or company are often obligated to assign ownership of their patents to that institution as stipulated by agreements signed when first joining the institution.

In order to be patented, an invention must meet three key requirements; in short, the invention must be novel, non-obvious, and useful. Utility patents may be granted for new and useful processes invented, design patents may be granted for a novel creation, and plant patents may be granted for new and asexually reproduced distinct plants (USPTO, 2023). Prior art, including existing written, oral, and in use information, is evaluated by the USPTO to determine the patentability of an invention based on novelty and a non-obvious inventive steps. Depending on the circumstances and timing, a public disclosure of an invention can dissatisfy the requirement of non-obviousness. The patent application filing date is the cutoff when public disclosures are no longer considered prior art. An alternative to filing a patent application is called defensive publication, where one establishes the invention as prior art in the public domain in order to prevent others from obtaining a patent on the invention.

A patent is comprised of the specification, which broadly includes background, description, scope, and embodiments of the invention, as well as one or more claims that define the scope of protection should the patent be granted. The patent application process consists of multiple phases where patent criteria, prior art, and other factors are examined and prosecuted. The process can be lengthy, costly, and result in changes to the claims of the patent application before it may be approved.

START-UPS

Starting a company may be a potential avenue to commercialize your scientific discoveries at some point during your career. There are many much more detailed and thorough resources entirely focused on this topic. In short, there are as many legal hazards as excitement in starting a company. Disagreements between individuals regarding control, money, or other business decisions arise all too frequently and devolve into legal disputes. The best advice we can provide in this brief note on the topic is to consult legal counsel as early as possible – at least well before money is involved. Seek advice and identify a trustworthy lawyer. Do not assume that a lawyer who works for your institution or for your startup company also represents your individual interests, unless you make a specific arrangement to that effect, sign a retainer agreement documenting this, and exchange some form of currency. An attorney working for a start-up may appropriately prioritize the company's interest over the specific interests of founders or other individuals associated with the company. When investing in such a high stakes effort, it is important to have someone looking out for your best interest. Starting a company can be a tremendously positive experience and exciting opportunity to implement your research in the real world. To avoid all too common legal issues, make sure that all legal and financial matters are clearly documented in writing and in consultation with legal experts, no matter how well you know your partners.

REFERENCES

Murphy, A., M. Stramiello, S. Lewis, & T. Irving. "Introduction to Intellectual Property: A U.S. Perspective." *Cold Spring Harbor Perspectives in Medicine* 5, no. 8 (2015): a020776. 10.1101/cshperspect.a020776

USPTO. (2023). *General Information Concerning Patents.* United States Patent and Trademark Office. https://www.uspto.gov/patents/basics/general-information-patents

21 Women in Science

Brett Goldsmith and Kiana Aran

THIS IS ABOUT EVERYBODY

Perhaps you've heard of the "leaky pipeline" model for training people in science and engineering. In this model, there is a pipeline of talent flowing from primary school through to success as a funded, tenured professor. At various stages of education and professional development, some of that talent leaks out of the pipeline, leaving fewer women and minorities on track to that professor position.

The "leaks" in the pipeline start early. There are many more girls held out of primary school than boys, and the way scientists are portrayed in media helps children define who a scientist is and whether they might include themselves in that definition. In countries with gender parity in primary school, there is no statistical difference in performance of girls and boys in science education.

Although the leaks start early, more women than men earn science and engineering associate's degrees in the United States, and more women than men earn science and engineering bachelor's degrees in the United States. This is one of the few leaks in the pipeline that is easy to fix: get girls in school and they will do well.

However, less than 30% of women are in science and engineering jobs in the United States (Figure 21.1). Women who are in those jobs do well. This is the other end of the pipeline: women who submit a follow-on grant application after already receiving a major research grant have statistically the same chance as men to have success in that follow-on application. This comes from a report put together by the National Science Foundation (NSF) in the United States.

The NSF defines a "scientist or engineer" as someone with a bachelor's degree or higher in any science or engineering discipline or someone who is employed in science, engineering, or a related position. Using this definition, women make up more than 50% of working scientists and engineers in the United States (Figure 21.1). The jobs scientists and engineers perform are split into categories of jobs in science and engineering, jobs related to science and engineering, and jobs not related to science and engineering (Figure 21.2). While women are underrepresented among science and engineering jobs, women find more positions in the other two categories, filling 57.7% of the jobs related to science and engineering and 54.4% of the jobs not related to science (but held by a scientist or engineer).

The definitions of these groups of jobs are important. Notably, "jobs related to science" include healthcare positions, where women represent a clear majority of the workforce (Figure 21.3). "Jobs not related to science" include most jobs in industry, including high-paying jobs in sales and management.

 DOI: 10.1201/9781003301400-21

USA, Total STEM Workforce   USA, Science Jobs

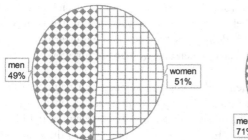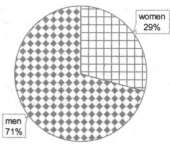

FIGURE 21.1 (Left) Distribution between men and women in the total STEM workforce in the United States. (Right) Distribution between men and women in "science and engineering" jobs in the United States. (National Center for Science and Engineering Statistics, 2023.)

Job Categories and Percentage of Jobs for Scientists and Engineers

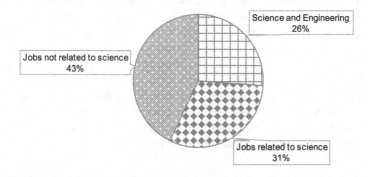

FIGURE 21.2 The relative percentage of jobs worked by scientists and engineers in the categories of science and engineering jobs, jobs related to science, and jobs not related to science. Only 26% of scientists and engineers work in jobs classified as "science and engineering."

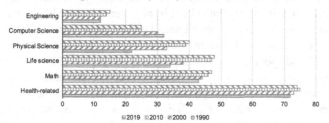

FIGURE 21.3 Underrepresentation of women is a significant problem in some technical fields such as engineering, computer science, and physical science. During the past 30 years, women's participation in computer science has decreased. Women's high presence in health-related occupations is largely a result of nursing and the continuing influence of traditional gender roles (Fry et al., 2021).

An important tool for personally overcoming unfair circumstances in your career is understanding your identity as a scientist or engineer. Do you identify as a scientist or engineer? Why? Does your definition of "scientist or engineer" match what society expects? It may be helpful to know that only 26% of that total STEM workforce fills jobs in the category defined as "science and engineering." A large majority of us in science and engineering, men and women, do not work in "science and engineering."

The lack of parity between men and women in science and engineering jobs indicates a complex combination of a lack of opportunity, lack of encouragement, unfair cultural pressure, and lack of diverse leadership. Each scientific field has its own set of challenges in this regard. Each type of job has its own different set of challenges, too. Each woman is a unique person with her own history, goals, and experiences. No individual person can make parity happen.

This dilemma is really about whether the current landscape of opportunities and rewards in science is a pure meritocracy. Science is not a pure meritocracy. That affects all of us negatively. First, each of us will be faced with unfair circumstances in our career. We can focus on women because it's trivial to see that there should be statistical parity between women and men, so it is easy to identify areas of improvement. The features that drive the underrepresentation of women in certain technical fields largely overlap with the features that drive underrepresentation of other groups, such as men of color, people whose parents did not go to college, and people who have a socioeconomically disadvantaged background (National Academies of Science, Engineering, and Medicine, 2020). Second, there are and will be labor shortages in certain disciplines such as computer science that can be significantly addressed by increasing representation of women in these fields. Third, the lack of parity between women and men in science is a moral and ethical issue within the scientific community that should be addressed.

WHY SO MANY OF US ARE IN FIELDS "OUTSIDE" OF SCIENCE

One of the key concepts in the "leaky pipeline" model of women in science is that women are more likely than men to leave their fields of scientific or engineering specialization as they progress in their career. While women and men are near parity at the undergraduate education level, men are more likely than women to receive a doctorate in most technical fields, and men are more likely than women to get a postdoc position across all technical fields. That trend continues with the percentage of women relative to men decreasing for assistant professor positions, associate professor positions, and full professor positions. This has been going on long enough that it is not only an issue of waiting for today's undergraduate students to work their way through the system and become professors; women do actually leave science careers at a higher rate than men as they progress further along the academic career track. This reflects the concept of persistence: the proportion of a particular group of people working in science who are still in their field after a number of years. Women in physics, chemistry, and biology do not persist in those fields over the typical academic career trajectory. However, the persistence of women in engineering and computer science is quite good. Women represent a low percentage (about 20%) of the bachelor's degree earners in engineering and represent the same percentage of professors in those fields (National Academies of Science, Engineering, and Medicine, 2020).

Over the last 40 years, academics and government officials have undertaken several efforts to increase the proportion of women in science. In part, this led to parity in the number of women receiving bachelor's degrees in science, and parity in the number of PhDs in some fields such as the biological sciences. We've already mentioned that women who receive their first research grant and apply for a follow-on grant have an equal chance to receive that follow-on award as a male researcher. There have also been significant improvements in the way faculty are hired, and many departments make a great effort to avoid unintentional bias in the faculty hiring process. However, getting to the point of being confident in landing a faculty position is very difficult, and more men than women apply for faculty positions, even in fields where there is gender parity in granted PhDs (National Research Council, 2010).

Why does this happen?

There are several critical junctures in preparing for a career in science, many little decisions and circumstances that can lead to an accelerated trajectory or a barrier that must be overcome. Some of these critical junctures are obtaining key work experience as an undergraduate, graduate school admission, recruitment into leading laboratories once in graduate school, assignment to key projects once in a laboratory, gaining a prestigious postdoctoral position, and assignment of tasks during a postdoc likely to prepare someone for a faculty application.

Several studies have been run to test whether there is an effective bias favoring men over women in these type of early career positions. In one study, faculty members in physics, chemistry, and biology were asked to evaluate a resume that was identical other than the gender. Across all fields, the male resume was rated higher than the female resume. This was true even when looking at how the faculty who were women rated the resume. This reflects that even in scientific fields where women have achieved parity, there are still unique biases and pressures that women face (Moss-Racusin et al., 2012).

The various technical fields of science and engineering provide unique challenges. For example, physics has a history and culture of "immortals" such as Newton and Einstein. These field-defining giants delineate the career model and promote a sense of individual scientific heroism. This impresses upon the young physicist that success requires immediate raw talent. A poor personal comparison to Einstein may be harder to shake off by a woman who is wondering if she belongs in physics.

Biology, on the other hand, is culturally understood to require long hours of effort. The stories of the scientists who define the prototypical biology career are more modern and deal with struggles largely tackled as part of a team. However, this cultural difference doesn't necessarily mean that everything is easy for women working in biology. The emphasis on "the grind" over inspiration in biology results in a culture of long hours and blurred work-life balance.

The challenges raised by childcare and family planning across science are real and disproportionately fall on women. To participate in scientific leadership requires many years of training that typically overlaps with the period of life when non-scientists start a family. In science, the frequent expectation is that each of us will prioritize training and scientific work over any personal interests. This is not easy for anyone and is particularly difficult for women. Many academic institutions have improved childcare considerations for faculty who are women, but there remain extraordinary financial and time pressures on graduate students and postdocs. In addition, the cultural stereotypes that place responsibility for nurturing on women may lead to a "cultural tax" of extra responsibilities on women in the scientific workplace.

There are personality traits that are present in both men and women, but which are stereotypically seen as male (assertiveness, ambition, competitiveness) and are generally valued in scientific culture. When women project these traits, they are often criticized. The emphasis, generally, on these stereotypically male traits in scientific culture can signal that women do not belong in science and may contribute to this unintentional bias and generally discourage women to continue in science careers (National Academies of Science, Engineering, and Medicine, 2020).

This idea of stereotypical traits can also be applied to goals, how we talk about science, and what we do with the science we work on. From a social psychology point of view, both men and women share a high degree of regard for goals related to personal agency and achievement, such as directing our own research and being recognized in our fields. However, women tend to put a higher regard than men on goals that are communal, such as solving a problem for society or another person. Although science has a very communal goal-oriented role within society, scientific culture generally emphasizes hierarchy and the importance of achieving personal agency (control over research direction) as the ultimate career goal. This is thought to be a driving reason for the difficulty science faces in achieving gender parity at a senior leadership level compared to other traditionally male-dominated fields with a more communal culture, such as law, business, and medicine (Diekman and Steinberg, 2013).

A Harvard Business School study showed that women tend to associate high-level positions with conflict. Unlike men, women tend to believe that taking a leadership position will require them to compromise their other important life goals (Gino and Brooks, 2015). This may mean that some women who excel at science find more professional fulfillment working in what we have defined as jobs related to science (*i.e.*, medicine) and jobs not related to science (*i.e.*, business) rather than jobs in science.

WHAT WE CAN DO

If you're a woman early in your career in science …

1. *Define your own identity and career path*

One of the features that most correlates with persistence in a science career is self-identifying as a scientist or engineer. Do you identify as a scientist or engineer? What does that mean to you? There are many scientists and engineers who are women, who have come before, and who have accomplished great things, from

Marie Curie to Rosalind Franklin to Angela Merkel. Find a role model that inspires you. The important thing is that you feel professionally engaged and excited to forge or follow a career path.

2. Develop leadership skills

This is more than managing an intern. There are classes, programs, and mentors available through organizations like Athena and Society of Women Engineers that help develop skills and experience in mastering workplace leadership and conflict resolution. Very rarely is there any formal management training in science, so a little bit of formal training in this area can help you stand above the crowd. These kinds of certifications, and the skills that come with them, can make a big difference in both helping you feel more confident in achieving your career goals as well as projecting to potential employers that you intend to be in or develop into a leadership role.

3. Avoid imposter syndrome

There are many of us in science who look at the people who are "big successes" in our scientific fields and feel like there is no way we could ever be like those people. Many times, we look at those who are simply "doing well" in our own labs and feel like we're not measuring up. If you find that you avoid volunteering for the most challenging projects, are afraid to ask for help, or feel like you have to work extra to keep up with your peers, you may be dealing with something often called "imposter syndrome." Overcoming this is both extraordinarily difficult and simple: one must internally acknowledge one's own accomplishments and believe that they do have a productive role to play in science.

There are a few steps you can take to avoid or reduce how this impacts you. The primary goal here is to reduce the need for external validation and build habits that encourage self-validation. Identify what is in your control and change it.

- Identify negative thoughts like "everyone else always gets ..." and "I never achieve..." Fact check yourself when you see this and actively look for facts that challenge this thinking.
- Celebrate your achievements and remind yourself of them when you have doubts.
- Find small wins every day.
- Keep reminders around of your achievements.

4. Develop your team

Find a mentor or sponsor who can help you define and move through your career path (read the next section to find out what you should ask of your mentor). Identify and start working with allies and people who have professional interests that parallel yours (read the last section to find out how they should work with you).

If you're a mentor …

1. Adequate mentoring has been shown to address the gender imbalance in science. If you're a mentor to women (or really anyone), the first thing to do is to understand the responsibilities of mentorship. A common guide for mentors in higher education defines a few principles that should guide mentorship (Johnson, 2015):
   - *Beneficence:* Promote your mentee's best professional interests.
   - *Nonmaleficence:* Avoid using the mentor role for harm.
   - *Autonomy:* Avoid promoting dependency vs. independence.
   - *Fidelity:* A sense of loyalty between mentor and mentee.
   - *Fairness:* Equal treatment.
   - *Privacy:* Avoid revealing sensitive material without consent.
2. Ease orientation of young researchers into the scientific career system. Help them identify their individual career goals and the steps that will be required to reach them. Mediate important contacts with experienced researchers and help them build their professional network (Barabino et al., 2019).
3. To pursue a career in science, it is helpful to have role models. History provides plenty of great women scientists. You likely know more about some of these people than your students. You likely know some women who are currently working in science who can be regarded as role models. Talking about role models can be helpful (Barabino et al., 2019).
4. Moving young women scientists out of their cultural comfort zone can play a positive role in helping those young women understand and internalize their value and capabilities. A study through the Erasmus program (European Community Action Scheme for the Mobility of University Students) showed that women who went through a period of study that put them in a different environment, culture, and language perform better when they return to their local university (Barabino et al., 2019).
5. Create a positive work environment. This is a critical element to achieving gender parity. Scientists in general require inclusion, full participation, and respect to perform their best. These conditions are easier to achieve among the majority demographic groups in the workplace. Evaluation of whether a work environment is positive comes from the workers. Ask them about it even if you think everything is great. Create transparent organizational structures that promote fairness. Emphasize both personal and professional respect as part of the work environment (Barabino et al., 2019).
6. Create and enforce a set of standard working hours that are consistent with a family-friendly workplace.
7. Be an ally as well as a mentor.

If you're an ally (that's everyone) …

1. Sponsorship can be thought of as the "next level" of mentorship and has a much larger impact on developing leaders than mentorship and coaching

alone. Perhaps you can sponsor someone; perhaps you can connect someone to a sponsor. Sponsors advocate publicly for advancement of individual people from a position of power and influence. A sponsor may be someone like a dean, a grant manager, or an industry executive. Sponsorship can help women overcome a common lack of visibility within an organization and ensure that they have access to high-profile projects that pave the way for future leadership positions (National Academies of Science, Engineering, and Medicine, 2020).

2. Find ways to recognize and reward outstanding contributions. There are many award programs available to people at all levels of science. Look for institutional awards for teaching, service, or exemplary performance. Women are underrepresented in many honorary societies. Be proactive in identifying awards that women can and should apply for, and encourage women to apply (National Academies of Science, Engineering, and Medicine, 2020).

3. Use language that signals to women that they belong in your workplace. Use gender neutral terms (*e.g.*, "folks" versus "guys"). Focus on inclusion of and respect for the women in your workplace.

4. When involved in hiring, keep the job requirements focused, and remove requirements for personality traits that may be stereotypically male (*e.g.*, "thrives in a competitive environment"). Understand and record the evaluation criteria ahead of time. If cultural fit is an important part of your hiring process, understand exactly what cultural fit means ahead of time. Cultural fit can be a very positive tool but is sometimes used as an excuse to select only people with similar backgrounds. Avoid generalizations and focus on "individualizing" each applicant. If you're in a male-dominated workplace, before starting any evaluations of applicants, visualize how an effective, powerful woman would work well with your team. This kind of positive visualization helps overcome unintentional bias. Avoid often illegal questions about personal life such as marriage and children (National Academies of Science, Engineering, and Medicine, 2020).

5. Foster a positive work environment. In addition to what is covered under the "mentor" section above, understand the particular challenges that women face in science. It's been well researched that the traits seen as positive in men (ambition, self-promotion, competitiveness, assertiveness) are treated as negative traits in women (this is fairly universal; both men and women will make this same evaluation). Similarly, an impression of competence in science is often mutually exclusive with an impression of warmth and likeability. Understand these trends. Work with individuals in a fair and equal manner, identifying and countering any unintentional gender bias. Don't feel guilty that you have the same gender biases as everyone else (men and women), but do feel responsible for overcoming them. Help those around you to do the same.

CONCLUSION

Women make up half (or more) of the scientific and engineering workforce yet are spread very unevenly across job categories. This is clearly a problem for women pursuing careers in science and engineering. This is also a problem for all of us in science, as it indicates that there is a combination of cultural expectation, institutional practice, and unintentional bias that is preventing a group of people from participating and contributing fully to a field that stands to benefit greatly from their involvement. We have presented some easy actions that each of us can take to make our scientific work environments more welcoming to women. This is not a new issue, and it has been examined and studied extensively. What we've covered here is just an introduction. These principles can be applied to other under-represented minority groups as well. The interested reader (prospective student, job applicant, admissions officer, hiring manager, mentor, ally) should continue to learn and seek further opportunities for growth and improvement.

REFERENCES

Barabino, G., M. Frize, F. Ibrahim, E. Kaldoudi, L. Lhotska, L. Marcu, ... E. Bezak. "Solutions to Gender Balance in STEM Fields through Support, Training, Education and Mentoring: Report of the International Women in Medical Physics and Biomedical Engineering Task Group." *Science and Engineering Ethics* 26 (2019): 275–292.

Diekman, A., and M. Steinberg. "Navigating Social Roles in Pursuit of Important Goals: A Communal Goal Congruity Account of STEM Pursuits." *Social and Personality Psychology Compass* 7, no. 7 (2013): 487–501.

Fry, R., B. Kennedy, and C. Funk. *STEM Jobs See Uneven Progress in Increasing Gender, Racial and Ethnic Diversity*. Washington, DC: Pew Research Center, 2021.

Gino, F., and A. W. Brooks. *Explaining Gender Differences at the Top*. Boston: Harvard Business Review, 2015.

Johnson, W. B. *On Being a Mentor: A Guide for Higher Education Faculty*. New York: Routledge, 2015.

Moss-Racusin, C. A., J. F. Dovidio, V. L. Brescoll, M. J. Graham, and J. Handelsman. "Science Faculty's Subtle Gender Biases Favor Male Students." *PNAS* 109 (2012): 16474–16479.

National Academies of Science, Engineering, and Medicine. *Promising Practices for Addressing the Underrepresentation of Women in Science, Engineering, and Medicine: Opening Doors*. Washington, DC: The National Academies Press, 2020.

National Center for Science and Engineering Statistics. *Diversity and STEM*. Washington, DC: National Science Foundation, 2023.

National Research Council. *Gender Differences at Critical Transitions in the Careers of Science, Engineering, and Mathematics Faculty*. Washington, DC: The National Academies Press, 2010.

22 Scientific Career Trajectories

Nir Qvit and Samuel J. S. Rubin

INTRODUCTION

At the beginning of their careers, scientists tend to focus on developing a solid foundation of knowledge and skills that will serve them well throughout the rest of their careers. Often, the first step of the process is to obtain a bachelor's degree in a relevant field, followed by further specialization through advanced degrees, such as master's and/or doctoral degrees. During this sequence, students are required to engage in coursework, lab work, and collaborative projects that will help them gain expert knowledge in one or more fields of study.

Despite the fact that doctoral programs and postdoctoral positions provide essential training and specialization for scientists to carry out independent research, they do not necessarily prepare them for such an endeavor. Advanced proficiency and accomplishment also require skills related to project management, business acumen, and communication, among others.

After completing undergraduate and graduate studies, scientists typically enter the next phase of their careers in postdoctoral research or early-career positions. Traditionally, postdoctoral researchers worked in academia under the guidance of established scientists and further honed their skills while conducting supervised semi-independent research and developing an independent research program with which to apply for faculty positions. However, today scientists can also perform postdoctoral research in industry or choose to obtain industry jobs following a postdoc in academia.

In Chapter 17, we discussed pursuing a scientific career in industry or academia. In this chapter, we evaluate other career options you might want to consider too. Graduate and postdoctoral researchers may choose academic or industry positions, while others explore alternative career paths as product managers, consultants, science policymakers, and entrepreneurs.

ALTERNATE CAREER PATHS

There is a wide range of training pathways depending on the specific field of study and the individual's career goals. We provide a brief overview of the most common degrees and career paths. Although the following description provides a general overview, keep in mind that there are many variations and alternative routes within each trajectory.

DOI: 10.1201/9781003301400-22

Bachelor's degrees are the first step in many career paths and typically take three to four years to complete, depending on the field and country. During this period, students acquire foundational knowledge in their chosen field and develop critical thinking, research, and communication skills.

Upon graduating from a bachelor's program, some students may choose to work as research assistants (RAs) or technicians to gain practical experience. Conducting experiments, collecting and analyzing data, and providing assistance to more experienced researchers are some of the activities they perform in these entry-level positions. As a result of this hands-on experience, individuals gain a deeper understanding of their field and build a solid foundation for further studies.

For those who wish to gain advanced knowledge and specialize in a particular field, a master's degree (MS/MA) is an option. A master's degree program typically lasts one to three years and involves coursework, research, and a thesis or capstone project (see Chapter 10 for a more in-depth discussion of master's degrees). By earning a master's, students are able to gain deeper theoretical knowledge as well as practical skills, enabling them to pursue advanced roles or go on to doctoral studies.

The Doctor of Philosophy (PhD) degree prepares students for careers in research, academia, industry, or other specialized areas. The duration of a doctoral program is usually four to six years depending on the field and type of program, and it requires coursework, rigorous in depth research, independent study, and writing and defending a dissertation. A PhD candidate contributes original research, advances knowledge, and makes significant contributions to the field in which he or she studies.

Those who wish to practice medicine and/or conduct medical research may pursue a Doctor of Medicine (MD) degree. Typically, medical school requires four years of intensive study following a bachelor's degree. To develop knowledge of human health and disease, diagnosis, treatment, and clinical acumen, both class-room instruction and clinical rotations are included in the curriculum. Upon graduation, aspiring physicians undertake specialty residency training in their chosen specialty, and some go on to complete subspecialty fellowships.

The MD/PhD dual-degree program offers a comprehensive path for those who are interested in combining clinical practice with scientific research. Through this path, medical education is integrated with scientific research, enabling individuals to pursue careers as physicians and scientists. An MD/PhD program usually takes seven to eight years to complete at minimum. As part of this process, one generally completes two years of medical school classroom-based learning, followed by PhD courses and dissertation research, and finally the ultimate two clinical years of medical school. Advantages of the combined MD/PhD program include a unified admission process for both degrees, an accelerated timeline, and funded medical education (similar to most PhDs, in contrast to paid tuition for standard MD programs), while disadvantages include competitiveness of the programs and interruptions in clinical training to pursue research.

SCIENCE CAREERS IN THE UNITED STATES RELATIVE TO OTHER COUNTRIES

Science training and careers differ between the United States and many other countries. In the United States, formal education is highly emphasized, with extensive graduate research opportunities. Alternatively, international careers may be influenced by a variety of educational systems and can vary significantly depending on the country and its educational system. Funding mechanisms also differ, with some countries relying more or less on public sources to support scientific research. United States federal funding agencies (*e.g.*, the National Institutes of Health (NIH) and the National Science Foundation (NSF)) provide significant support for scientific research. Some other countries offer substantial government grants and fellowships, while others rely on philanthropic organizations or industry partnerships.

In the United States, most aspiring scientists pursue undergraduate degrees in biology, chemistry, physics, or engineering, which provide a broad foundation in scientific principles and methodologies. Next, individuals often attend graduate school, which offers extensive research opportunities. This allows students to delve deeper into their chosen field. During these research-intensive programs, individuals are prepared for a career in academia, industry, or government research.

On the international front, training paths for science careers vary significantly depending on the country and its educational system. In some countries, students pursue undergraduate degrees followed by graduate studies, similar to the United States. However, there are countries where specialized undergraduate programs are tailored to specific scientific fields. Typically, these programs combine rigorous coursework with hands-on laboratory experiences, providing students with a solid theoretical and practical foundation.

PRODUCT MANAGEMENT

A product manager is responsible for overseeing the success of a product as well as leading the interdisciplinary team responsible for its continuous improvement over time. Through their involvement in research and development, collaboration, networking, and commercialization, scientific product managers contribute to growth, impact, and advancement of scholarly scientific knowledge. Product managers in the sciences are responsible for assessing promising areas of study or "unmet needs" in order to guide the development of new products as well. The first steps in this process involve defining research objectives, allocating resources, and providing a clear strategy for research. The manager plays a crucial role in the communication between scientists and industry partners as well as with other stakeholders. Creating a culture of collaboration, partnerships, and knowledge-sharing allows product managers to support scientists in expanding their scientific networks and diversifying their careers. Furthermore, product managers also play an important role in the process of commercializing a product. To bring revolutionary products to the market successfully, it is vital to understand both the scientific and business aspects of the space. Product managers assess market potential, identify the target market, develop marketing strategies, and transform research findings into commercially viable products. Thus, scientific training and expertise allow individuals to assume product management roles in scientific fields that require specialized knowledge to fulfill their roles.

CONSULTING

Scientists can also pursue careers in consulting. Consultants provide expertise and guidance by applying their scientific knowledge and analytical skills to new problems and scenarios. Depending on your experience, you may be able to work independently or in a consulting firm, providing services such as data analysis, market research, technology assessments, or strategic planning to clients. Consultants in scientific fields work with clients across a wide range of industries, including healthcare, pharmaceuticals, energy, environmental sciences, technology, and even academia. A scientific or research consultant can be hired to assist with the evaluation of research proposals. This can include conducting feasibility studies, evaluating the impact of evolving technologies, or assisting with decision-making processes. As a consultant, it is essential that you have the ability to communicate effectively, solve problems, and translate complex scientific concepts into actionable recommendations.

SCIENCE COMMUNICATION

Science communication strives to establish a bridge between scientific research and society as a means of promoting understanding, appreciation, and the use of scientific information to make informed decisions. In order to spread the word (often to a lay audience) about the important work of scientists, science writers may publish articles or blogs, contribute to television programming or lectures, create podcasts, etc. For effective science communication, one must convey complex concepts in a manner that is clear, engaging, and relatable to the audience. The language and presentation style scientists use and the way they explain their research to audiences differs depending on who they are addressing, and scientists should adjust their language and presentation style accordingly. By mastering communication skills and making use of multiple platforms, scientists can enhance the visibility and impact of their research. In turn, this will contribute to the advancement of knowledge and society as a whole. Scientists who work for science communication organizations (often journals and media outlets) contribute to the dissemination of scientific information by translating complex scientific concepts into accessible language, developing educational materials, organizing public events, and engaging with a wide range of audiences.

A medical writer is a specialized form of science communication that focuses on creating accurate and clear documents about patient health and disease. Medical writers are responsible for translating complicated medical and scientific information into clear, understandable language that can be communicated to a variety of audiences, including healthcare professionals, researchers, regulatory agencies, patients, and caregivers. During the past few years, medical writers have played an important role in the pharmaceutical, biotechnology, and healthcare industries. These professionals are responsible for compiling, analyzing, and presenting scientific data in a concise and coherent manner that is often also accessible to lay audiences. They are in close contact with researchers, clinicians, and regulatory experts, making sure that the content is evidence-based, easily understandable, and tailored to the target audience. Types of medical writing include research papers, clinical trial protocols, regulatory submissions, healthcare education materials, and patient information leaflets, among many others. In addition, many medical writers focus on clinical study protocols, drug safety reports, and submission dossiers that are required by regulatory authorities, such as the Food and Drug Administration (FDA) or the European Medicines Agency (EMA). Medical writers contribute significantly to the medical and healthcare fields by combining their expertise in science with their competence in writing.

GOVERNMENT AGENCIES

Scientific careers in government agencies offer opportunities for scientists to contribute to public policy and regulatory frameworks. Specializations include environmental policy, public health, energy, agriculture, and many others. Focusing on one of these areas allows scientists to gain a deeper understanding, participate in nuanced policy development, and offer scientific input in areas that are directly related to their technical expertise. An individual who demonstrates expertise, leadership qualities, and the ability to bridge the gap between science and policy can progress to management positions or become a scientific advisor to a high-ranking official, which may lead to significant impacts on policy. In addressing pressing challenges, such as public health crises, climate change, or technological advancements, these individuals have the opportunity to shape society for the better. Given the frequent disconnect between politics and science, there is a need for scientists who work for government agencies to balance scientific objectivity with policy considerations. While their primary focus must be to provide accurate and unbiased scientific information, they must also consider how their recommendations will affect political, social, and economic behavior. This balance can only be achieved by effective communication, collaboration with policymakers, and a thorough understanding of the process through which government agencies make their decisions.

Like other sectors, government scientists are often constrained by limited funding and resources, which makes their work difficult to carry out. Scientists at these agencies may have to deal with limited budgets, limited research opportunities, and bureaucratic processes. On the other hand, research scientists employed by NIH research institutes or other specific agencies may have unusual access to research funding and unparalleled resources, making these attractive research positions.

NON-PROFIT ORGANIZATIONS

Scientific careers in non-profit organizations offer the opportunity for scientists to contribute their expertise for a specific cause or the greater good. Non-profit organizations may hire scientists to develop innovative solutions and initiatives to benefit communities or the environment. Scientific careers in non-profit organizations require a combination of scientific expertise, communication skills, and a strong commitment to the organization's mission. Scientists should possess a solid foundation in their respective scientific disciplines and demonstrate the ability to apply their knowledge to address real-world challenges. Effective communication, collaboration, leadership, and project management skills are also essential for success in non-profit roles. In these roles, scientists may address a wide range of problems, such as those related to public health, climate change, biodiversity conservation, and social justice.

SUMMARY

Scientific careers are multifaceted and can evolve across different sectors. Funding, job availability, and personal interests will affect individual paths. Navigating the trajectory of a scientific career requires adaptability, resilience, and the willingness to grab opportunities as they arise. Whether in academia, industry, government, non-profit organizations, or as a consultant, scientists can play key roles in advancing knowledge, addressing societal challenges, and shaping the future through their research and technical skills.

REFERENCES

Agarwal, Rajshree, and Atsushi Ohyama. "Industry or Academia, Basic or Applied? Career Choices and Earnings Trajectories of Scientists." *Management Science* 59, no. 4 (April 1, 2013): 950–970.

Cech, Erin A., and Mary Blair-Loy. "The Changing Career Trajectories of New Parents in Stem." *Proceedings of the National Academy of Sciences* 116, no. 10 (2019): 4182–4187.

Krogh, Lars Brian, and Hanne Moeller Andersen. "'Actually, I May Be Clever Enough to Do It.' Using Identity as a Lens to Investigate Students' Trajectories Towards Science and University." *Research in Science Education* 43, no. 2 (April 1, 2013): 711–731.

Larose, Simon, Catherine F. Ratelle, Frédéric Guay, Caroline Senécal, and Marylou Harvey. "Trajectories of Science Self-Efficacy Beliefs during the College Transition and Academic and Vocational Adjustment in Science and Technology Programs." *Educational Research and Evaluation* 12, no. 4 (August 1, 2006): 373–393.

Yaffe, Kirsten, Carol Bender, and Lee Sechrest. "How Does Undergraduate Research Experience Impact Career Trajectories and Level of Career Satisfaction: A Comparative Survey." *Journal of College Science Teaching* 44, no. 1 (2014): 25–33.

23 Epilogue

Nir Qvit and Samuel J. S. Rubin

The purpose of this book is to provide a valuable resource to those who are seeking a creative and research-oriented career in science and/or technology, ranging from academics to industry research and development, start-up businesses, non-profit organizations, and government. Career success can be achieved in many ways. Throughout the course of this book, we discussed a variety of topics, dilemmas, and challenges faced by young scientists as they embark on their careers. This book presents challenges and not guarantees and although broad is by no means entirely comprehensive. To achieve long-term success and satisfaction in a research career, you will have to develop and apply healthy habits over an extended period of time. We hope that this book has provided more encouragement than pause by preparing those interested in science. A career in science offers a wide range of opportunities for professional development, exploration, and the possibility of making a meaningful contribution to society.

Scientists play crucial roles in advancing knowledge and innovation through research. Even though it is important to have strong individual expertise, networking and collaboration are equally essential. A multidisciplinary and increasingly interconnected world requires scientists to build strong networks and embrace a teamwork mentality both within and outside their immediate environment (e.g., lab, department, university, field of study, etc.). Developing effective networking skills is a vital skill as you pursue your professional goals. Collaboration promotes synergy and facilitates the pooling of different expertise across many disciplines. This culture is fostered by holding regular team meetings, brainstorming sessions, and sharing resources, which help to promote creativity and accelerate the progress of research. A surprising number of valuable connections and serendipitous discoveries can be made as a result of informal interactions, such as coffee breaks or social events among scientists. The exchange of ideas stimulates innovation and facilitates the development of novel approaches and solutions. Besides providing professional opportunities, networking can also lead to career advancement and access to funding opportunities. Building a robust network increases visibility and professional credibility. Last but not least, networking fosters a sense of community by providing support and mentoring from peers and experts to nurture a sense of belonging. Networking and teamwork are integral components of a successful scientific career and cannot be overstated.

To ensure your success as a scientist, it is crucial to plan ahead, as science is progressing at an unprecedented rate. A solid foundation for scientific progress can be created by establishing a strategic budget, assembling powerful teams, and

DOI: 10.1201/9781003301400-23

creating a robust infrastructure through which scientists are able to optimize their resources and build a solid body of research. It is essential for the advancement of science that we have an efficient and well-equipped infrastructure.

After careful planning, scientists should follow their passion and embrace the unexpected. This allows for the discovery of groundbreaking insights and innovation through curiosity. A scientist who has a passion for what he or she does is more likely to invest his or her time, energy, and resources in exploring uncharted territory. Those who are passionate about their research demonstrate a keen sense of curiosity, perseverance, and dedication that is evident in their work. These individuals are more equipped to overcome challenges and expand knowledge.

Failure is embraced by scientists as a necessary part of the research process and should thus be considered a vital stepping stone toward success. As a result of research mistakes, scientists are forced to reevaluate their assumptions, refine their methodologies, and develop alternative approaches to conduct their experiments more successfully.

Enthusiasm for research often leads to innovative approaches and novel methodologies. A scientist who is deeply invested in their field is more likely to think creatively, question established norms, and come up with unconventional hypotheses. This type of mindset fosters transformational breakthroughs that have revolutionized entire fields of science. Over the course of history, there have been numerous instances when groundbreaking discoveries were made unintentionally. Serendipity has played an important role in the advancement of science for generations. Accidental discoveries are often the result of careful observations, flexibility, and an open-minded approach to research.

Among the most notable examples of serendipitous discoveries is Alexander Fleming's accidental discovery of penicillin's antibacterial properties in 1928. Fleming observed that a mold called *Penicillium* inhibited bacterial growth in his Petri dishes while studying bacteria. It was this serendipitous encounter that led to the development of the first antibiotic, which revolutionized medicine and saved countless lives. In the process of scientific exploration, mistakes are a natural part of the process. These situations often result in valuable opportunities for educational growth as well as unexpected breakthroughs in science and technology.

There are many topics covered in this book. We encourage readers to explore topics and revisit them as they become relevant in the future. Seek advice from mentors and colleagues and continue to update your sources of information, enhance your skillset, and grow your network.

As you embark on your scientific journey, we extend our best wishes for a fulfilling and successful career. May your pursuit of knowledge and contribution to your field lead to groundbreaking discoveries and advancements that benefit society as a whole. Allow your passion for science to guide you to new heights as you embrace challenges, find joy, and embrace discovery.

We would highly appreciate your feedback on how this book has influenced your scientific journey. Whether it provides insights, inspiration, or a fresh perspective, your experiences and reflections are useful for shaping future editions. This will help us better serve aspiring scientists like you. Please feel free to share your thoughts and experiences with us so that we can continue to refine and enhance our offerings.

Index

Page numbers in *italics* refer to figures and those in **bold** refer to tables.

Printed in the United States
by Baker & Taylor Publisher Services